SCHOLAR Study Guide

CfE Advanced Higher Physics
Unit 1: Rotational Motion and Astrophysics

Authored by:
Julie Boyle (St Columba's School)

Reviewed by:
Grant McAllister (Bell Baxter High School)

Previously authored by:
Andrew Tookey (Heriot-Watt University)

Campbell White (Tynecastle High School)

Heriot-Watt University

Edinburgh EH14 4AS, United Kingdom.

First published 2015 by Heriot-Watt University.

This edition published in 2015 by Heriot-Watt University SCHOLAR.

Distributed by the SCHOLAR Forum.

SCHOLAR Study Guide Unit 1: CfE Advanced Higher Physics

1. CfE Advanced Higher Physics Course Code: C757 77

ISBN 978-1-911057-00-0

Printed and bound by CPI Group (UK) Ltd, Croydon, CR0 4YY

Acknowledgements

Thanks are due to the members of Heriot-Watt University's SCHOLAR team who planned and created these materials, and to the many colleagues who reviewed the content.

We would like to acknowledge the assistance of the education authorities, colleges, teachers and students who contributed to the SCHOLAR programme and who evaluated these materials.

Grateful acknowledgement is made for permission to use the following material in the SCHOLAR programme:

The Scottish Qualifications Authority for permission to use Past Papers assessments.

The Scottish Government for financial support.

The content of this Study Guide is aligned to the Scottish Qualifications Authority (SQA) curriculum.

Contents

Topic 1

Kinematic relationships

Contents

Prerequisite knowledge

- *Knowledge of the difference between vector and scalar quantities.*

- *Calculus - differentiation and integration*

- *Some familiarity with the kinematic relationships would be useful.*

- *Understand the shape of displacement-time graphs, velocity-time graphs and acceleration-time graphs for constant acceleration.*

Learning objectives

By the end of this topic you should be able to:

- *use calculus notation to represent velocity as the rate of change of displacement with respect to time;*

- *use calculus notation to represent acceleration as the rate of change of velocity with respect to time;*

- *use calculus notation to represent acceleration as the second differential of displacement with respect to time;*

- *derive the three kinematic relationships from the calculus definitions of acceleration and velocity:*

$$v = u + at \qquad\qquad s = ut + \tfrac{1}{2}at^2 \qquad\qquad v^2 = u^2 + 2as;$$

- *apply these equations to describe the motion of a particle with uniform acceleration moving in a straight line;*

- *state that the gradient of a graph can be found by differentiation;*

- *state that the gradient of a displacement-time graph at a given time is the instantaneous velocity;*

- *state that the gradient of a velocity-time graph at a given time is the instantaneous acceleration;*

- *state that the area under a graph can be found by integration;*

- *state that for a velocity-time graph the displacement can be found by integrating between the limits;*

- *state that for an acceleration-time graph the change in velocity can be found by integrating between the limits;*

- *use calculus methods to solve problems based on objects moving in a straight line with varying acceleration.*

1.1 Introduction

This topic deals with linear motion. From the Higher course, you should already be familiar with the vector nature of the quantities displacement, velocity and acceleration. You may also recall the kinematic relationships describing the motion of objects with constant acceleration. We will now use calculus to derive these relationships, starting from the definitions of instantaneous acceleration and velocity.

We will then look at some examples of the use of motion sensors to study uniform acceleration in a straight line. We will also explore graphical methods for analysing motion, for both uniform and varying acceleration. Lastly, we will finish the topic by studying some specific examples of objects moving in a straight line with varying acceleration.

1.2 Calculus methods

The average velocity of an object can be found from the equation:

$$v_{\text{average}} = \frac{\Delta s}{\Delta t}$$

where Δs is the change in displacement and Δt is the change in time. However, if we want to find the instantaneous velocity, we must consider a very small time interval. In other words, we must find the derivative of displacement with respect to time. So we need to use calculus notation:

$$v = lim\frac{\Delta s}{\Delta t} \qquad (\text{as } \Delta t \to 0)$$
$$v = \frac{ds}{dt}$$

Likewise, we could find the average acceleration at Higher by using the equation:

$$a_{\text{average}} = \frac{\Delta v}{\Delta t}$$

but to find instantaneous acceleration we will need to consider the rate of change of velocity with respect to time. In other words, we must differentiate velocity with respect to time.
This gives:

$$a = lim\frac{\Delta v}{\Delta t} \qquad (\text{as } \Delta t \to 0)$$
$$a = \frac{dv}{dt}$$

Since $v = \frac{ds}{dt}$, the acceleration can also be written as:

$$a = \frac{d}{dt}\frac{ds}{dt} = \frac{d^2s}{dt^2}$$

This means that acceleration can either be expressed as the first derivative of velocity with respect to time or as the second derivative of displacement with respect to time:

$$a = \frac{dv}{dt} \quad \text{or} \quad a = \frac{d^2s}{dt^2}$$

It is worth noting that acceleration is a vector quantity and so an object will accelerate if either the magnitude of its velocity changes or the direction alters. In this topic we will solely explore objects moving in a straight line, but the next topic will consider the acceleration of an object moving in a circular path.

1.3   Deriving the equations for uniform acceleration

The equations describing motion with a constant acceleration can be derived from the definitions of acceleration and velocity.

We will start with the definition of instantaneous acceleration: Acceleration is the rate of change of velocity.

$$a = \frac{dv}{dt}$$

We are looking at motion where a is a constant. To find the velocity after time t, we will integrate this expression over the time interval from $t = 0$ to $t = t$.

$$\int_u^v dv = \int_{t=0}^t a\,dt = a\int_{t=0}^t dt$$

Carrying out this integration

$$[v]_u^v = a\,[t]_0^t$$
$$\therefore v - u = at$$
$$\therefore v = u + at$$

$$(1.1)$$

. . . .

Equation 1.1 gives us the velocity v after time t, in terms of the acceleration a and the initial velocity u. Velocity is defined as the rate of change of displacement. We will now use this definition to derive the second of the kinematic relationships.

$$v = \frac{\mathrm{d}s}{\mathrm{d}t}$$

Since $v = \frac{ds}{dt}$ and $v = u + at$, we have:

$$\frac{\mathrm{d}s}{\mathrm{d}t} = u + at$$

Integrating over the time interval from 0 to t gives:

$$\int_{s=0}^{s} \mathrm{d}s = \int_{t=0}^{t} (u + at)\mathrm{d}t$$

$$\therefore [s]_0^s = \left[ut + \frac{1}{2}at^2 \right]_0^t \qquad (1.2)$$

$$\therefore s = ut + \frac{1}{2}at^2$$

. . . .

Equation 1.2 gives us the displacement s after time t, in terms of the acceleration and the initial velocity. To obtain the third kinematic relationship, we first rearrange Equation 1.1.

$$t = \frac{v - u}{a}$$

Substituting this expression for t into Equation 1.2 gives us

$$s = u \left(\frac{v - u}{a} \right) + \frac{1}{2}a \left(\frac{v - u}{a} \right)^2$$

$$\therefore 2as = 2vu - 2u^2 + v^2 - 2vu + u^2$$

$$\therefore 2as = -2u^2 + v^2 + u^2$$

$$\therefore v^2 = u^2 + 2as$$

$$(1.3)$$

. . . .

We have obtained three equations relating displacement s, time elapsed t, acceleration a and the initial and final velocities u and v. Note that with the exception of t, these are all vector quantities.

$$v = u + at \qquad\qquad s = ut + \frac{1}{2}at^2 \qquad\qquad v^2 = u^2 + 2as$$

Although we are dealing with vector quantities, we are only considering the special case of motion in a straight line. It is vital to ensure we assign the correct +ve or -ve sign to each of the quantities s, u, v and a.

It is also useful to remember that for an object moving in a straight line with uniform acceleration, the average velocity over a period of time is given by $\frac{(u+v)}{2}$.

The displacement can therefore be found from $s = \frac{(u+v)}{2} \times t$.

1.4   Uniform acceleration in a straight line

We are now going to use the kinematic relationships to describe the motion of a particle moving in one dimension.

1.4.1   Horizontal motion

The kinematic relationships can be used to solve problems of motion in one dimension with constant acceleration. In all cases we will be ignoring the effects of air resistance.

Example

A car accelerates from rest at a rate of 4.0 m s^{-2}.

1. What is its velocity after 10 s?

2. How long does it take to travel 72 m?

3. How far has it travelled after 8.0 s?

We can list the data given to us in the question

u = 0 m s^{-1} (the car starts from rest)
a = 4.0 m s^{-2}

1. We are told that the time elapsed t = 10 s and we wish to find v. So with u, a and t known and v unknown, we will use the equation $v = u + at$.

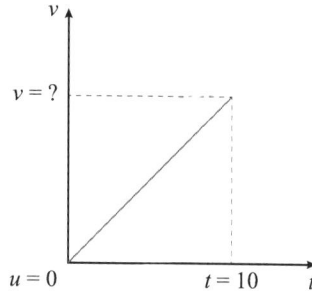

$$v = u + at$$
$$\therefore v = 0 + (4.0 \times 10)$$
$$\therefore v = 40 \text{ m s}^{-1}$$

2. In part 2 we know u, a and s, and t is the unknown, so we use $s = ut + \frac{1}{2}at^2$.

$u = 0$ m s^{-1}
$a = 4.0$ m s^{-2}
$s = 72$ m
$t = ?$

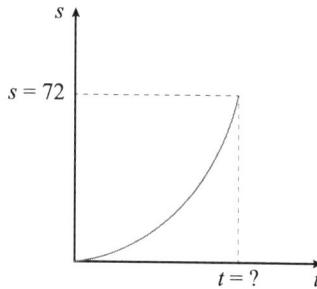

$$s = ut + \frac{1}{2}at^2$$
$$\therefore 72 - 0 + \left(\frac{1}{2} \times 4.0 \times t^2\right)$$
$$\therefore 72 = 2t^2$$
$$\therefore t^2 = 36$$
$$\therefore t = 6.0 \text{ s}$$

3. Finally in part 3 we are asked to find s, given u, a and t, so we will again use $s = ut + \frac{1}{2}at^2$, this time using different data.

$u = 0$ m s^{-1}
$a = 4.0$ m s^{-2}
$t = 8.0$ s
$s = ?$

$$s = ut + \tfrac{1}{2}at^2$$
$$\therefore s = 0 + \left(\tfrac{1}{2} \times 4 \times 8^2\right)$$
$$\therefore s = 128\,\text{m}$$

. . . .

The same strategy should be used in solving all of the problems for constant acceleration. Firstly, list the data given to you in the question. This will ensure that you use the correct values when you perform any calculations, and should also make it clear to you which of the kinematic relationships to use. It is often useful to make a sketch diagram with arrows, to ensure that any vector quantities are being measured in the correct direction.

Horizontal motion

Go online

Suppose a car is being driven at a velocity of 12.0 ms^{-1} towards a set of traffic lights, which are changing to red. The car driver applies her brakes when the car is 30.0 m from the stop line. What is the minimum uniform deceleration needed to ensure the car stops at the line?

There is an online activity available which will provide further practice in this type of problem.

. . . .

1.4.2   Vertical motion

When dealing with freely-falling bodies on Earth, the acceleration of the body is the acceleration due to gravity, $g = 9.8$ m s^{-2}. If any force other than gravity is acting in the vertical plane, the body is no longer in free-fall, and the acceleration will take a different value. Problems should be solved using exactly the same method we used to solve horizontal motion problems.

Example

A student drops a stone from a second floor window, 15 m above the ground.

1. How long does it take for the stone to reach the ground?

2. With what velocity does it hit the ground?

When dealing with motion under gravity, we must take care with the direction we choose as the positive direction. Here, if we take a as a positive acceleration, then v and s will also be positive in the downward direction.

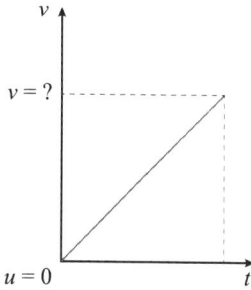

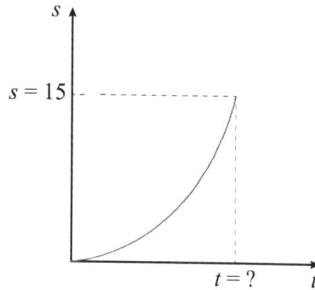

We are told that
$u = 0$ m s^{-1}
$a = g = 9.8$ m s^{-2}
$s = 15$ m

1. To find t, we use $s = ut + \frac{1}{2}at^2$.
   Putting in the appropriate values

$$s = ut + \tfrac{1}{2}at^2$$
$$\therefore 15 = 0 + \left(\tfrac{1}{2} \times 9.8 \times t^2\right)$$
$$\therefore 15 = 4.9t^2$$
$$\therefore t^2 = \frac{15}{4.9}$$
$$\therefore t^2 = 3.06$$
$$\therefore t = 1.7\,\text{s}$$

2. To find v, we use $v^2 = u^2 + 2as$.
   Again, we put the appropriate values into this kinematic relationship

$$v^2 = u^2 + 2as$$
$$\therefore v^2 = 0 + (2 \times 9.8 \times 15)$$
$$\therefore v^2 = 294$$
$$\therefore v = 17\,\text{m s}^{-1}$$

. . . .

Difficulties sometimes occur when the initial velocity is directed upwards, for example if an object is being thrown upwards from the ground. To solve such a problem, it is usual to take the vertical displacement as being positive in the upwards direction. The velocity vector is then positive when the object is travelling upwards, and negative when it is returning to the ground. In this situation the acceleration is always negative, as it is always directed towards the ground.

Quiz: Motion in one dimension

Go online

Useful data:

Gravitational acceleration on Earth g	$9.8\ m\ s^{-2}$

Q1: An object is moving with a uniform acceleration of 5 m s^{-2}. A displacement-time graph showing the motion of this object has a gradient which

a) increases with time.
b) decreases with time.
c) equals 5 m s^{-1}.
d) equals 5 m s^{-2}.
e) equals 0.

. . . .

Q2: A car accelerates from rest with a uniform acceleration of 2.50 m s^{-2}. How far does it travel in 6.00 s?

a) 7.50 m
b) 15.0 m
c) 18.75 m
d) 45.0 m
e) 112.5 m

. . . .

Q3: Neglecting air resistance, a stone dropped from the top of a building 125 m high hits the ground after

a) 1.25 s.
b) 4.00 s.
c) 5.05 s.
d) 12.7 s.
e) 25.5 s.

. . . .

Q4: A diver jumps upwards with an initial vertical velocity of 3.00 m s^{-1} from a diving board which is 8.00 m above the swimming pool. With what vertical velocity does he enter the pool?

a) 3.0 m s^{-1}
b) 3.6 m s^{-1}
c) 9.0 m s^{-1}
d) 13 m s^{-1}
e) 24 m s^{-1}

. . . .

Q5: A stone is dropped from a window 28.0 m above the ground. What is the velocity of the stone when it is 10.0 m above the ground?

a) 10.0 m s^{-1}
b) 14.0 m s^{-1}
c) 18.8 m s^{-1}
d) 23.4 m s^{-1}
e) 98.1 m s^{-1}

. . . .

1.5 Graphical methods for uniform acceleration

We will look at how motion with constant acceleration can be represented in graphical form. We can use graphs to show how the acceleration, velocity and displacement of an object vary with time.

1.5.1 Graphs for motion with constant velocity

The graphs below all represent motion with a constant velocity.

Figure 1.1: Graphs for motion with constant velocity

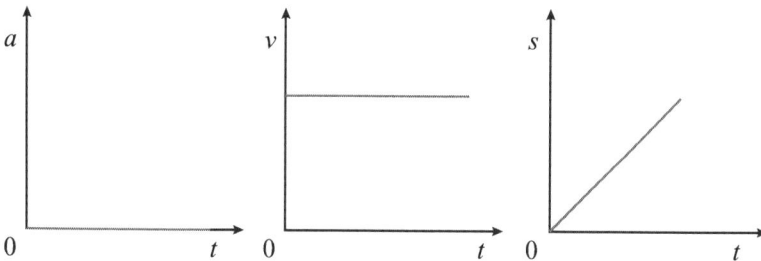

. . . .

By inspecting the gradient of the displacement-time graph, we can see that the steepness of the graph is constant. This means that $\frac{ds}{dt}$, the instantaneous velocity, is constant.

The velocity-time graph is horizontal and so its gradient io zeiu. This tells us that $\frac{dv}{dt}$, the acceleration, is zero.

You will recall from Higher that acceleration can be found from a velocity-time graph by calculating the gradient of the line on the graph. The gradient of a straight line can be found from the equation:

$$\text{gradient} = \frac{y_2 - y_1}{x_2 - x_1}$$

For uniform acceleration this is equivalent to using the equation:

$$a = \frac{v - u}{t}$$

We will further explore the gradient method in section 1.7. There, we will look at varying acceleration, where the equation $a = \frac{v-u}{t}$ will no longer be valid.

You will also recall from Higher that the area under a velocity time graph is equal to the displacement. Similarly, it is worth noting that the area under an acceleration time graph is equal to the change in velocity.

Graphs for motion with constant velocity

There is an online activity at this stage displaying more information on the graphs for motion with constant velocity.

Go online

. . . .

1.5.2   Graphs for motion with constant positive acceleration

The graphs below all represent motion with a constant positive acceleration.

Figure 1.2: Graphs for motion with constant positive acceleration

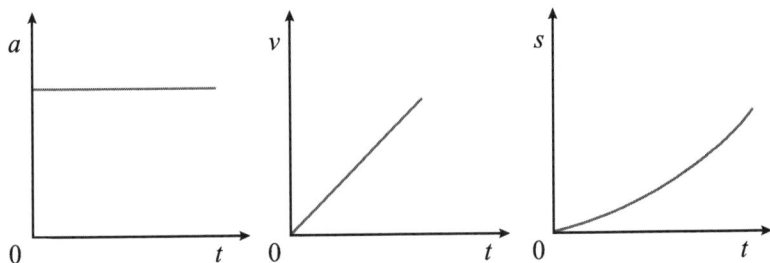

. . . .

By inspecting the gradient of the displacement-time graph, we can see that the graph is getting steeper. This means that $\frac{ds}{dt}$, the instantaneous velocity, is increasing.

By inspecting the gradient of the velocity-time graph, we can see that the steepness of the graph is constant. This means that $\frac{dv}{dt}$, the acceleration, is constant.

Note that since acceleration and velocity are the same sign, the object is speeding up.

Graphs for motion with constant positive acceleration

There is an online activity at this stage displaying more information on the graphs for motion with constant positive acceleration.

Go online

....

1.5.3   Graphs for motion with constant negative acceleration

The graphs below all represent motion with a constant negative acceleration.

Figure 1.3: Graphs for motion with constant negative acceleration

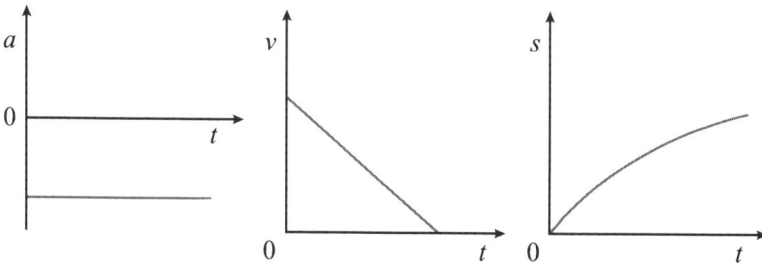

....

By inspecting the gradient of the displacement-time graph, we can see that the graph is getting less steep. This means that $\frac{ds}{dt}$, the instantaneous velocity, is decreasing.

By inspecting the gradient of the velocity-time graph, we can see that the steepness of the graph is constant and negative. This means that $\frac{dv}{dt}$, the acceleration, is constant and negative.

Note that since the acceleration and the velocity are the opposite sign, the object is slowing down.

Graphs for motion with constant negative acceleration

There is an online activity at this stage displaying more information on the graphs for motion with constant negative acceleration.

Go online

....

1.5.4   Example graphs

Examples

1.

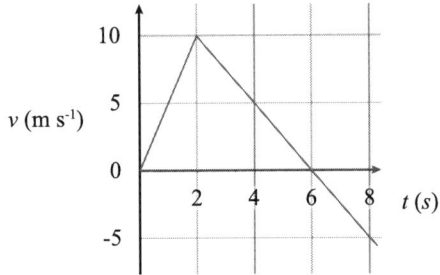

For the velocity - time graph shown, calculate:

1. The acceleration from 2.0 s to 8.0 s.
2. The displacement after 8.0 s.

1.

acceleration = gradient of v-t graph = -15/6 = -2.5 m s^{-2}

or

$$a = \frac{v - u}{t}$$
$$a = \frac{-5 - 10}{6}$$
$$a = -2.5 \text{ m s}^{-2}$$

2. displacement = area under v-t graph = $1/2$ x 6 x 10 - $1/2$ x 2 x 5 = 25 m

. . . .

2.

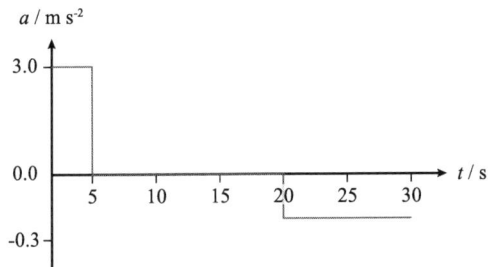

Find the change in velocity in:

1. The first 5 seconds.
2. Between 0 and 30 seconds.

1.

change in velocity $= $ area under $a - t$ graph

$=3 \times 5$

$=15$ m s^{-1}

2.

change in velocity $=$ area under $a - t$ graph

$=(3 \times 5) - (0.3 \times 10)$

$=15 - 3$

$=12$ m s^{-1}

. . . .

Interactive example graphs

There is an online activity at this stage exploring these relationships.

. . . .

Go online

1.5.5   Objects in freefall

The displacement of an object with constant acceleration can be found from

$$s = ut + \frac{1}{2}at^2$$

For a dropped object, the initial velocity is 0 m s^{-1} and the acceleration is the acceleration due to gravity, g. So the value of g can be evaluated from the gradient of the straight line graph of s against t^2.

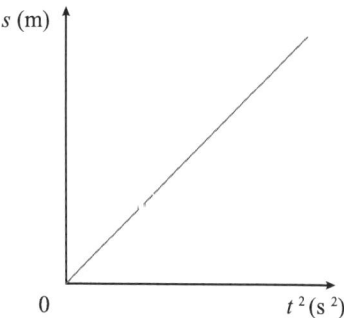

$$s = \frac{1}{2}gt^2$$

A graph of s against t^2 has gradient $\frac{1}{2}g$.

The acceleration due to gravity can also be found by considering the displacement-time graph. The gradient of the $s - t$ graph is increasing since the instantaneous velocity increases. The instantaneous velocity, v, at a given time, t, can be found by drawing a tangent to the graph and determining its gradient. To sketch a tangent to the graph, a straight line must be drawn such that it touches the curve only at one point.

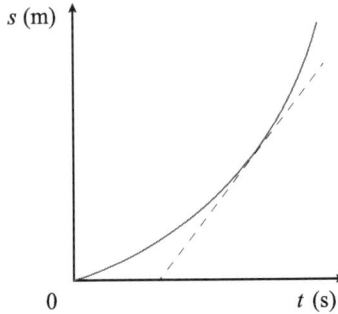

The acceleration due to gravity can then be found by substituting the values of v and t into the equation

$$a = \frac{v - u}{t}$$

1.6   Motion sensors

The displacement of an object can be measured using a motion sensor.

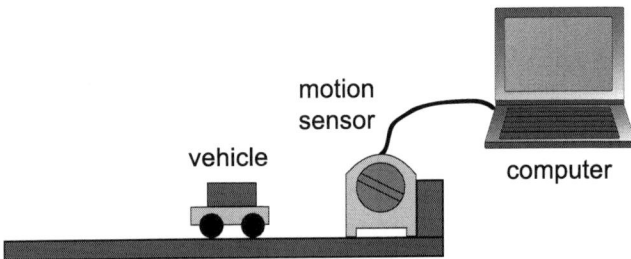

The motion sensor contains an ultrasound transmitter. This sends out pulses of high frequency sound that are reflected back to a special microphone on the sensor. The sensor measures the time for the sound pulses to return to the microphone and uses the speed of sound to calculate the distance of the object from the sensor. This is usually displayed on a computer screen as a displacement time graph.

Light gates are devices that can be used to measure how long it takes an object to pass a point.

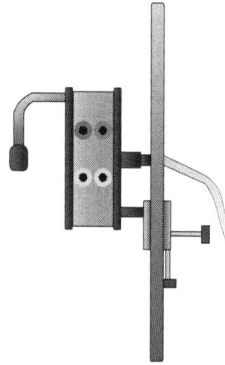

A beam of light passes from the box to a sensor on the arm of the light gate. The light gate is attached to a timer such as a datalogger. When an object breaks the beam the timer measures the amount of time the beam is blocked for. If the length of the object is known then the speed of the object can be calculated, this is often done by the timing device. The following set up can be used to measuring the acceleration of a vehicle on a slope.

datalogger

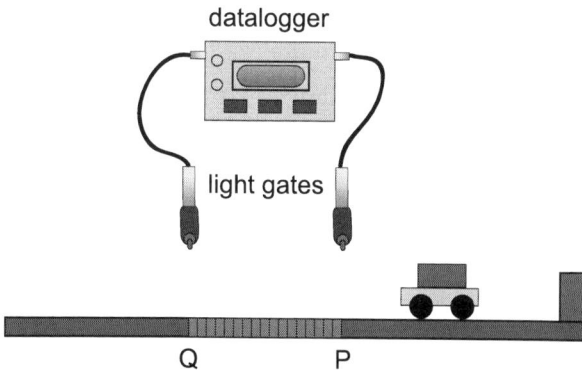

light gates

Q   P

Two light gates are placed at P and Q and connected to the datalogger. The length of the car is measured an entered into the datalogger. When the car passes through the light gate at P the datalogger measures the time it takes to pass and then calculates its speed by dividing the length of the car by the time taken, this is the initial velocity, u. It then measures the time taken for the car to go from P to Q, t. Finally it measures the time taken to pass through the light gate at Q and then calculates its speed by dividing the length of the car by the time taken, this is the final velocity, v. The acceleration of the car can be calculated using equation $a = \frac{v-u}{t}$. The apparatus shown above can also be used to calculate the acceleration by another method.

If instead of measuring the time taken to go from light gate P to Q the distance between the two light gates, s, is measured then equation $v^2 = u^2 + 2as$ can be used to calculate the acceleration.

Acceleration can be measured with a single light gate as long as a 'double mask' is used.

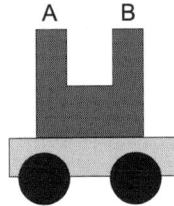

The above mask is used as follows:
When the first section of the mask, A, passes through the light gate the datalogger measures the time to pass and then calculates the initial speed, u, by dividing the length of A by the time. The time for the gap to pass through the light gate, t, is recorded. When the final section, B, passes through the light gate the datalogger measures the time to pass and then calculates the final speed, v, by dividing the length of B by the time. The acceleration of the light gate can be calculated using equation $a = \frac{v-u}{t}$.

The motion of an object can also be analysed using video techniques. This has become increasingly important in the film and computer areas of the entertainment industry. The illustration below shows how motion capture is used to analyse the movement of a person.

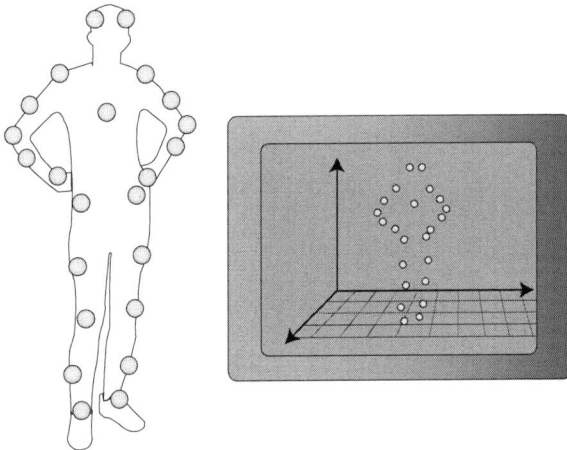

The person has a number of lights placed on their clothes and then stand against a dark background containing a grid. A video is taken of them as they move across the grid and then analysed frame by frame so that their motion can be reproduced in computer graphics. Similar techniques are used when analysing the motion of vehicles in crash test laboratories.

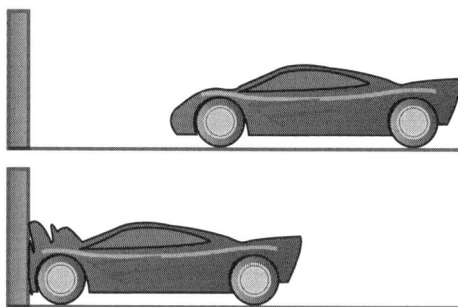

The diagram shows two still frames from the video of a car being crashed at high speed. By making measurements from these frames computer software can be used to analyse the motion of the car. This will allow the acceleration of the car to be calculated and hence the forces acting on occupants of a car in a high speed crash.

1.6.1 Motion of bouncing ball

Motion of bouncing ball

There is an online activity exploring these relationships.

. . . .

Go online

1.6.2 Motion on a slope

Motion of toy car moving down a slope and hitting a stretched elastic band

This interactivity is available online only.

. . . .

Go online

1.7   Graphical methods for non-uniform acceleration

So far we have assumed uniform (constant) acceleration. In other words, the change in velocity with time has been constant. For non-uniform (varying) acceleration, the change of velocity with time is not constant throughout the motion. So the slope of the velocity-time graph will be changing and it may have curved sections. Therefore, to find the acceleration at a given point in time, you need to draw a tangent to the curve at that point and then calculate its gradient.

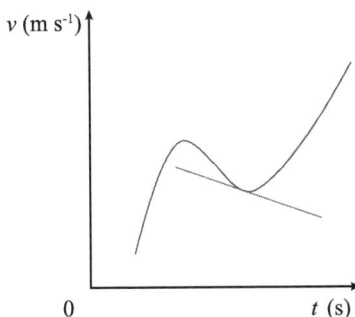

Velocity-time graph

There is an online activity at this stage displaying seven tangents.

. . . .

Similarly, the velocity at a given point in time can be found from a displacement-time graph by drawing a tangent to the curve at that point and calculating its gradient.

Displacement-time graph

There is an online activity at this stage displaying seven tangents.

. . . .

If an equation describing the velocity as a function of time is known, then the acceleration at a given time can be found by differentiating the equation and solving for that specific time. Derivatives are just formulae that let us find the slope at any point on a function. In fact, differentiating and finding the gradient of the tangent to the graph are really just the same mathematical process.

You should by now be familiar with the process of finding the displacement from a velocity-time graph by calculating the area under the graph. What you are really doing is integrating. Now let's consider what happens if the acceleration varies and the velocity-time graph is not linear. We will be unable to consider the area of simple shapes. Instead, we will need to calculate the displacement by integrating the velocity function between the limits. This means we must integrate and then substitute in the limits, subtracting the value at the lower limit from the value at the higher limit.

In a similar way, the change in velocity can be determined by finding the definite Integral of the acceleration function i.e. integrating with limits.

This can all be summarised as follows

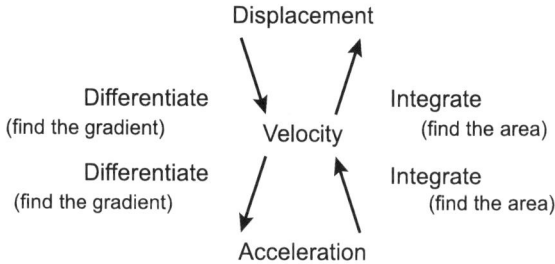

Displacement

Differentiate   Integrate
(find the gradient)   Velocity   (find the area)

Differentiate   Integrate
(find the gradient)   (find the area)

Acceleration

Transformations between graphs of motion

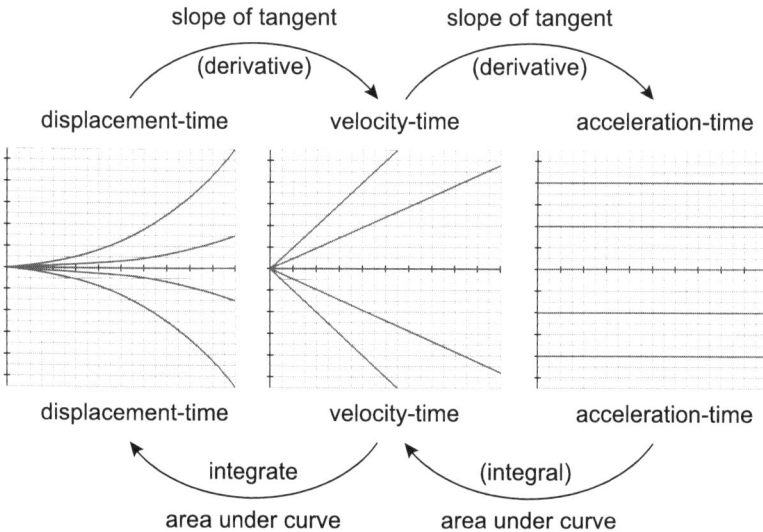

Go online

slope of tangent   slope of tangent
(derivative)   (derivative)
displacement-time   velocity-time   acceleration-time

displacement-time   velocity-time   acceleration-time

integrate   (integral)
area under curve   area under curve

There is an animation at this stage displaying the graphs of motion.

. . . .

1.8   Examples of varying acceleration

Now let's turn our attention to some specific examples of objects moving in a straight line with varying acceleration.

Examples

1.

The velocity in m s^{-1} of a moving particle is given by the equation

$$v = 4t^3 - 5t^2 + 6t$$

where t is the time in seconds.
Find the acceleration at the instant t = 0.5 seconds.

$$v = 4t^3 - 5t^2 + 6t$$
$$a = \frac{dv}{dt} = 12t^2 - 10t + 6$$
At $t = 0.5$ seconds
$$a = (12 \times 0.5^2) - (10 \times 0.5) + 6$$
$$a = 4 \text{ m s}^{-2}$$

. . . .

2.

The acceleration of a space shuttle in m s^{-2} is described by the expression

$$a = \left(7.0 \times 10^{-7}\right) t^3 - \left(3.9 \times 10^{-4}\right) t^2 + (0.11) t + 0.86$$

where t is the time in seconds.
Determine the change in velocity of the space shuttle throughout the first 100 seconds.

$$\frac{dv}{dt} = a$$
$$\int_u^v dv = \int_0^{100} \left(7.0 \times 10^{-7}\right) t^3 - \left(3.9 \times 10^{-4}\right) t^2 + (0.11) t + 0.86 \quad dt$$
$$[v]_u^v = \left[\frac{7.0 \times 10^{-7} t^4}{4} - \frac{3.9 \times 10^{-4} t^3}{3} + \frac{0.11 t^2}{2} + 0.86 t \right]_0^{100}$$

(Note the constant of integration was omitted because it would be in both parts. It always disappears when the parts are subtracted and so it is normally left out of the working when integrating with limits.)

$$v - u = \left(\frac{7.0 \times 10^{-7} (100)^4}{4} - \frac{3.9 \times 10^{-4} (100)^3}{3} + \frac{0.11 (100)^2}{2} + (0.86 \times 100) \right) - (0)$$
$$v = 17.5 - 130 + 550 + 86 - 0$$
$$v = 520 \text{ m s}^{-1}$$

. . . .

3.

The acceleration of a particle in m s^{-2} is described by the expression

$$a = 2.5t$$

where t is in seconds. The initial velocity of the particle is 6.0 m s^{-1}.
Determine its velocity after 2.0 seconds.

$$\frac{dv}{dt} = a$$

$$\int_6^v dv = \int_0^2 2.5t \quad dt$$

$$[v]_6^v = \left[\frac{2.5t^2}{2}\right]_0^2$$

$$v - 6 = \frac{2.5(2)^2}{2} - 0$$

$$v = 11 \text{ m s}^{-1}$$

. . . .

4.

The acceleration of a particle in m s^{-2} is described by the equation.

$$a = 5t$$

where t is in seconds. It starts from rest at $t = 0$ seconds.
Determine the time taken for the particle to accelerate to 8.0 m s^{-1}.

$$\frac{dv}{dt} = a$$

$$\int_0^8 dv = \int_0^t 5t \quad dt$$

$$[v]_0^8 = \left[\frac{5t^2}{2}\right]_0^t$$

$$8 = \frac{5t^2}{2} - 0$$

$$t = 1.8 \ s$$

. . . .

5.

The velocity of a car in m s⁻¹ over a period of 8 seconds is described by the equation

$$v = 2.5t - 0.2t^2$$

where t is the time in seconds.

Find the distance travelled in this time interval.

$$v = \frac{ds}{dt}$$

$$\int_0^s ds = \int_0^8 \left(2.5t - 0.2t^2\right) \; dt$$

$$[s]_0^s = \left[\frac{2.5t^2}{2} - \frac{0.2t^3}{3}\right]_0^8$$

$$s = \left[\left(\frac{2.5 \times 8^2}{2} - \frac{0.2 \times 8^3}{3}\right)\right] - [0]$$

$$s = 45.9 \text{ m}$$

. . . .

6.

When a car brakes to a halt, the displacement of the car from a reference point can be described by the equation

$$s = 5.8t - 2.5t^2$$

where t is the time in seconds.

Sketch a displacement-time graph for the motion.

$$v = \frac{ds}{dt} = \frac{d}{dt}\left(5.8t - 2.5t^2\right)$$

$$v = 5.8 - 5t$$

For $v = 0$ m s⁻¹, $t = \frac{5.8}{5} = 1.16 \; s$

So the car comes to a halt at 1.2 s. The corresponding displacement is as follows.

$$s = (5.8 \times 1.16) - \left(2.5 \times 1.16^2\right)$$

$$s = 3.4 \text{ m}$$

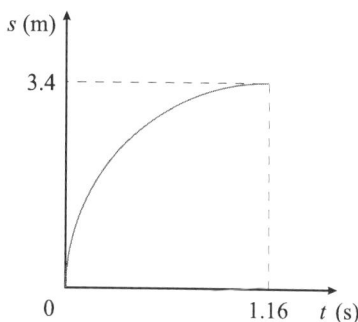

1.9 Extended information

Web links

There are web links available online exploring the subject further.

1.10 Summary

Summary

You should now be able to:

- use calculus notation to represent velocity as the rate of change of displacement with respect to time;

- use calculus notation to represent acceleration as the rate of change of velocity with respect to time;

- use calculus notation to represent acceleration as the second differential of displacement with respect to time;

- derive the three kinematic relationships from the calculus definitions of acceleration and velocity:

$$v = u + at \qquad s = ut + \tfrac{1}{2}at^2 \qquad v^2 = u^2 + 2as;$$

- apply these equations to describe the motion of a particle with uniform acceleration moving in a straight line;

- state that the gradient of a graph can be found by differentiation;

Summary continued

- state that the gradient of a displacement-time graph at a given time is the instantaneous velocity;

- state that the gradient of a velocity-time graph at a given time is the instantaneous acceleration;

- state that the area under a graph can be found by integration;

- state that for a velocity-time graph the displacement can be found by integrating between the limits;

- state that for an acceleration-time graph the change in velocity can be found by integrating between the limits;

- use calculus methods to solve problems based on objects moving in a straight line with varying acceleration.

1.11 Assessment

End of topic 1 test

Go online

The following test contains questions covering the work from this topic.

The following data should be used when required:

Gravitational acceleration on Earth g	$9.8\ m\ s^{-2}$

Q6:

A car accelerates uniformly from rest, travelling 55 m in the first 8.5 s.

Calculate its acceleration.

a = _____ m s^{-2}.

. . . .

Q7:

The brakes on a car can provide a deceleration of up to 4.9 m s^{-2}. The brakes are applied when the car is initially travelling at 20 m s^{-1}.

Calculate the minimum stopping distance.

Stopping distance = _____ m

. . . .

Q8:

A toy rocket is fired vertically upwards with an initial velocity of 16 m s^{-1}.

Calculate how long the rocket stays in the air before it returns to the ground.

---

t = _____ s

....

Q9:

A piano is being raised to the third floor of a building using a rope and pulley. The piano is 15 m above the ground, moving upwards at 0.25 m s $^{-1}$, when the rope snaps.

Calculate how much time elapses before the piano hits the ground.

t = _____ s

....

Q10:

The instantaneous acceleration of a body can be found by calculating the _____ of the tangent to the velocity-time graph.

....

Q11:

The displacement can be found from a velocity-time graph by determining the area under the graph. This process is equivalent to _____ between limits.

This process is equivalent to _____ between limits.

....

Q12:

The acceleration of a particle in m s^{-2} is described by the equation $a = 4.0t - 1.2$ where t is the time in seconds.

The particle starts from rest at the time $t = 0$ s.

Find the magnitude of its velocity at the instant $t = 5.0$.

v = _____ m s^{-1}

....

Q13:

The acceleration in m s^{-2} of an object can initially be described by the equation $a = 4.20t - 0.600$ where t is in seconds.

The object is already travelling at 5.00 m s^{-1} at $t=0$ s.

Find the magnitude of its velocity after 3.00 s.

v = _____ m s^{-1}

....

Q14:

The velocity of a particle moving in a straight line is given by the expression $v = 9.0t^2 - 0.5t$.

At time, $t=0$ s the displacement of the particle is 5.0 m.

Find the displacement of the particle at 2.0 s.

s = _____ m

....

Topic 2

Angular motion

Contents

Prerequisite knowledge

- *Kinematic relationships.*

- *Calculus - differentiation and integration.*

- *Addition and subtraction of vectors.*

- *Application of Newton's laws of motion to static and dynamic situations, including the use of free-body diagrams to solve problems.*

Learning objectives

By the end of this topic you should be able to:

- *measure angles and angular displacement in radians, and convert an angle measured in degrees into radians, and vice versa;*

- *use angular displacement, angular velocity and angular acceleration to describe motion in a circle;*

- *apply the equation $T = \frac{2\pi}{\omega}$ relating the periodic time to the angular velocity;*

- *derive the angular kinematic relationships, and apply them to solving problems of circular motion with uniform angular acceleration;*

- *relate the angular velocity to the tangential (linear) speed of a body moving in a circle, and derive the equation $v = r\omega$;*

- *apply the expression $a = r\alpha$ relating tangential and angular accelerations;*

- *derive the expressions for centripetal motion in terms of v and ω;*

- *calculate the centripetal acceleration and centripetal force of an object undergoing circular motion;*

- *use free-body diagrams to calculate centripetal forces;*

- *describe a number of real-life situations in which the centripetal force plays an important role.*

2.1 Introduction

In the Kinematics topic we studied the motion of objects travelling in a straight line. Now we will move on to study circular motion, which has a wide range of applications. These include the motion of planets around the Sun, the drum of a washing machine and a car taking a corner. We will consider the angular displacement, velocity and acceleration, and relate these quantities to the linear displacement travelled and the tangential speed and acceleration.

In the second part of this topic we will examine the centripetal force acting on an object. Without this force, an object would move off in a straight line at a tangent to the circle. We will then look at the variables that affect the size of the centripetal force acting on an object moving in a circular path.

2.2 Angular displacement and radians

In the first Topic of the course we investigated motion in a straight line. We are going to apply many of the ideas we met in that Topic to describe the motion of an object moving in a circle. We will see in the next Section that instead of using velocity and acceleration vectors, we will be using angular velocity and angular acceleration. Firstly we will look at angular displacement, which replaces the linear displacement we are used to dealing with. Imagine a disc spinning about a central axis, as shown in Figure 2.1. We can draw a reference line along the radius of the disc. The **angular displacement** after time t is the angle through which this line has swept in time t.

Figure 2.1: Angular displacement

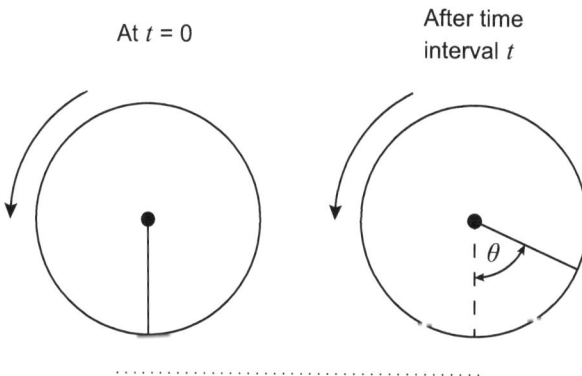

The angular displacement is given the symbol θ, and is measured in **radians** (rad). Throughout this Topic, radians will be used to measure angles and angular displacement. A brief explanation of radian measurement, and how radians and degrees are related, will be given next.

Figure 2.2: Radian measurement

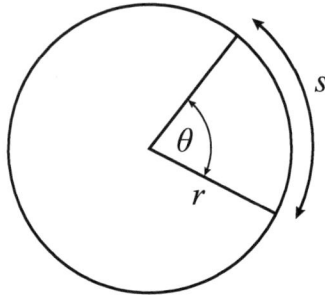

With reference to Figure 2.2, the angle θ, measured in radians, is equal to s/r. Since the radian is defined as being one distance (s) divided by another (r) then strictly speaking it is a dimensionless quantity, however the radian is regarded as a supplementary SI unit.

To compare radians with degrees, consider an angular displacement of one complete circle, equivalent to a rotation of 360°. In this case, the distance s in Figure 2.2 is equal to the circumference of the circle. Hence

$$\theta = \frac{s}{r}$$
$$\therefore \theta = \frac{2\pi r}{r}$$
$$\therefore \theta = 2\pi \text{ rad}$$

(2.1)

So 360° is equivalent to 2π rad, and this relationship can be used to convert from radians to degrees, and vice versa. It is useful to remember that π rad is equivalent to 180° and $\pi/2$ rad is equivalent to 90°. For the sake of neatness and clarity, it is common to leave an angle as a multiple of π rather than as a decimal, so the equivalent of 30° is usually expressed as $\pi/6$ rad rather than 0.524 rad.

Quiz: Radian measurement

Q1: Convert the angle 145° into radians.

Go online

a) 0.395 rad
b) 0.405 rad
c) 2.53 rad
d) 23.1 rad
e) 911 rad

....

Q2: What is the equivalent in degrees (°) to 1.20 radians?

a) 3.76°
b) 7.54°
c) 68.75°
d) 138°
e) 432°

....

Q3: Express 120° in radians.

a) $\pi/120$ rad
b) $\pi/4$ rad
c) $\pi/3$ rad
d) $2\pi/3$ rad
e) $3\pi/2$ rad

....

Q4: An object moves through 3 complete rotations about an axis. What is its total angular displacement?

a) $\pi/6$ rad
b) $\pi/3$ rad
c) 3π rad
d) 6π rad
e) $2\pi^3$ rad

....

Q5: An object moves 5.00 cm around the circumference of a circle of radius 24.0 cm. What is the angular displacement of the object?

a) 0.208 rad
b) 0.200π rad
c) 4.80 rad
d) 4.80π rad
e) 9.60π rad

....

2.3   Angular velocity and acceleration

Now that we have introduced the concept of angular displacement, it follows that the **angular velocity** ω is equal to the rate of change of angular displacement, just as the linear velocity v is the rate of change of linear displacement s.

$$\omega = \frac{d\theta}{dt} \qquad\qquad\qquad v = \frac{ds}{dt}$$

ω is measured in radians per second (rad/s or rad s^{-1}). The average angular velocity over a period of time t is the total angular displacement in time t divided by t.

Example

It takes the Moon 27.3 days to complete one orbit of the Earth. Assuming the Moon travels in a circular orbit at constant angular velocity, what is the angular velocity of the Moon?

The Moon moves through 2π rad in 27.3 days.

Now, 27.3 days is equal to   (27.3 × 24) = 655.2 hours
which is equal to   (655.2 × 60) = 39312 minutes
which is equal to   (39312 × 60) = 2358720 s

So the angular velocity ω is given by

$$\omega = \frac{\text{angular displacement}}{\text{time elapsed}}$$
$$\therefore \theta = \frac{2\pi}{2358720}$$
$$\therefore \theta = 2.66 \times 10^{-6} \text{ rad s}^{-1}$$

. . . .

You should also be aware of two other useful ways of describing the rate at which a body is moving in circular motion. One is to use the **periodic time** (or **period**) T, which is the time taken for one complete rotation. The other is to express the rate in terms of **revolutions per second**, which is the inverse of the periodic time. In the example above, the periodic time T for the Moon orbiting the Earth is 27.3 days, or 2.36×10^6 s in SI units. The rate of rotation is equal to $1/T = \frac{1}{2.36 \times 10^6} = 4.24 \times 10^{-7}$ revolutions per second.

The relationship between periodic time and angular velocity is

$$\omega = \frac{2\pi}{T}$$

(2.2)

. . . .

Having defined angular displacement θ and angular velocity ω, it should be clear that if ω is changing then we have an **angular acceleration**. The instantaneous angular acceleration α is the rate of change of angular velocity, measured in rad s^{-2}.

$$\alpha = \frac{d\omega}{dt}$$

(2.3)

. . . .

The average angular acceleration over time t is the total change in angular velocity divided by t.

Orbits of the planets

Following on from the above example, calculate the periodic times and angular velocities of the motion of Mercury, Venus and the Earth around the Sun.

Fill in the gaps in the table below.

Planet	Orbit radius (m)	Period (days)	Period (s)	Angular velocity (rad s^{-1})
Mercury	5.79×10^{10}	88.0		
Venus	1.08×10^{11}	225		
Earth	1.49×10^{11}	365		

. . . .

2.4   Kinematic relationships for angular motion

In the Kinematics topic we derived the kinematic relationships for linear motion with constant acceleration from the definitions of instantaneous acceleration and velocity. We can use the same technique to derive the equations of circular motion with constant angular acceleration by starting from the definitions of angular velocity and angular acceleration.

We begin with the definition of instantaneous angular acceleration

$$\alpha = \frac{d\omega}{dt}$$

Remember that we are only considering motion where α is a constant. Integrating the above expression over the time interval from 0 to t gives us

$$\int_{\omega_0}^{\omega} d\omega = \int_{t=0}^{t} \alpha dt = \alpha \int_{t=0}^{t} dt$$

Carrying out this integration

$$[\omega]_{\omega_0}^{\omega} = \alpha \, [t]_0^t$$
$$\therefore \omega - \omega_0 = \alpha t$$
$$\therefore \omega = \omega_0 + \alpha t$$

$$(2.4)$$

. . . .

Equation 2.4 gives us the angular velocity after time t in terms of the angular velocity at $t = 0$ and the angular acceleration. Now we can substitute for $\omega = {}^{d\theta}/_{dt}$ in this equation

$$\frac{d\theta}{dt} = \omega_0 + \alpha t$$

Integrating this equation over the same time interval

$$\int_{\theta=0}^{\theta} d\theta = \int_{t=0}^{t} (\omega_0 + \alpha t)$$

$$\therefore \theta = \left[\omega_0 t + \frac{1}{2}\alpha t^2 \right]_0^t$$

$$\therefore \theta = \omega_0 t + \frac{1}{2}\alpha t^2$$

(2.5)

. . . .

Equation 2.5 gives us the angular displacement after time t, again in terms of the angular velocity at $t = 0$ and the angular acceleration. Finally we use Equation 2.4, rearranged as follows

$$t = \frac{\omega - \omega_0}{\alpha}$$

Substituting this expression for t into Equation 2.5 gives us

$$\theta = \omega_0 \left(\frac{\omega - \omega_0}{\alpha} \right) + \frac{1}{2}\alpha \left(\frac{\omega - \omega_0}{\alpha} \right)^2$$

$$\therefore \alpha\theta = \omega_0 (\omega - \omega_0) + \frac{1}{2}(\omega - \omega_0)^2$$

$$\therefore 2\alpha\theta = 2\omega_0 (\omega - \omega_0) + (\omega - \omega_0)^2$$

$$\therefore 2\alpha\theta = 2\omega_0\omega - 2\omega_0^2 + \omega^2 - 2\omega_0\omega + \omega_0^2$$

$$\therefore 2\alpha\theta = -\omega_0^2 + \omega^2$$

$$\therefore \omega^2 = \omega_0^2 + 2\alpha\theta$$

(2.6)

. . . .

We now have a set of three kinematic relationships that describe circular motion with constant angular acceleration. If you have trouble remembering them, you should be able to work them out by comparison with their linear equivalents:

Linear motion	Circular motion
$v = u + at$	$\omega = \omega_0 + \alpha t$
$s = ut + \frac{1}{2}at^2$	$\theta = \omega_0 t + \frac{1}{2}\alpha t^2$
$v^2 = u^2 + 2as$	$\omega^2 = \omega_0{}^2 + 2\alpha\theta$

We can now use these equations to solve problems involving motion with constant angular acceleration.

Examples

1.

An electric fan has blades that rotate with angular velocity 80 rad s⁻¹. When the fan is switched off, the blades come to rest after 12 s. What is the angular deceleration of the fan blades?

We follow the same procedure as we used to solve problems in linear motion - list the data and select the appropriate kinematic relationship.

Here we are told
ω_0= 80 rad s⁻¹
ω = 0 rad s⁻¹
t = 12 s
α = ?

With ω_0, ω and t known and α unknown, we use $\omega = \omega_0 + \alpha t$ to find α.

$$\omega = \omega_0 + \alpha t$$
$$\therefore \omega - \omega_0 = \alpha t$$
$$\therefore \alpha = \frac{\omega - \omega_0}{t}$$
$$\therefore \alpha = \frac{0 - 80}{12}$$
$$\therefore \alpha = -6.7 \text{ rad s}^{-2}$$

So the angular acceleration is -6.7 rad s^{-2}, equivalent to a deceleration of 6.7 rad s^{-2}.

....

2.

A wheel is rotating at 35 rad s^{-1}. It undergoes a constant angular deceleration. After 9.5 seconds, the wheel has turned though an angle of 280 radians.
What is the angular deceleration?

$$\theta = \omega_0 t + \frac{1}{2}\alpha t^2$$

$$280 = (35 \times 9.5) + \left(\frac{1}{2} \times \alpha \times 9.5^2 \right)$$

$$\alpha = -1.2 \text{ rad s}^{-2}$$

(Please write unit as is already there for alpha, i.e. power notation)

So the angular acceleration is -1.2 rad s^{-2}, equivalent to a deceleration of 1.2 rad s^{-2}.

....

3.

An ice skater spins with an angular velocity of 15 rad s^{-1}. She decelerates to rest over a short period of time. Her angular displacement during this time is 14.1 rad.
Determine the time during which the ice skater decelerates.

$$\omega^2 = \omega_0^2 + 2\alpha\theta$$

$$0^2 = 15^2 + (2\alpha \times 14.1)$$

$$\alpha = -7.98 \text{ rad s}^{-2}$$

$$\omega = \omega_0 + \alpha t$$

$$0 = 15 - 7.98t$$

$$t = 1.9 \text{ s}$$

....

Quiz: Angular velocity and angular kinematic relationships

Q6: A compact disc is spinning at a rate of 8.00 revolutions per second. What is the angular velocity of the disc?

Go online

a) 0.020 rad s^{-1}
b) 0.785 rad s^{-1}
c) 1.27 rad s^{-1}
d) 25.1 rad s^{-1}
e) 50.3 rad s^{-1}

....

Q7: A turntable rotates through an angle of 1.8π rad in 4.0 s. What is the average angular velocity of the turntable?

a) 0.45π rad s^{-1}
b) 0.90π rad s^{-1}
c) 2.2π rad s^{-1}
d) 4.0π rad s^{-1}
e) 7.2π rad s^{-1}

. . . .

Q8: A disc is spinning about an axis through its centre with constant angular velocity 7.50 rad s^{-1}. What is the angular displacement of the disc in 8.00 s?

a) 0.938 rad
b) 1.07 rad
c) 9.55 rad
d) 60.0 rad
e) 377 rad

. . . .

Q9: At the top of a hill, the wheels of a bicycle are rotating at angular velocity 5.0 rad s^{-1}. When the cyclist reaches the bottom of the hill, the angular velocity has increased to 12 rad s^{-1}. If it took the cyclist 10 s to cycle down the hill, what was the angular acceleration of the wheels?

a) 0.35 rad s^{-2}
b) 0.70 rad s^{-2}
c) 1.4 rad s^{-2}
d) 1.7 rad s^{-2}
e) 5.95 rad s^{-2}

. . . .

Q10: A car parked on a slope begins to slowly roll forward. The angular acceleration of the wheels from rest is 0.20 rad s^{-2}. How long does it take for the wheels to rotate through one complete revolution?

a) 3.5 s
b) 5.0 s
c) 7.9 s
d) 12.6 s
e) 31 s

. . . .

2.5 Tangential speed and angular velocity

Suppose an object is moving in a circle of radius r with angular velocity ω. What is the relationship between ω and v, the velocity of the object measured in m s^{-1}?

We can answer this question by considering the way the radian is defined. Looking back to Figure 2.2, the angle θ, in radians, is given by

$$\theta = \frac{s}{r}$$

Differentiating this equation with respect to t gives us

$$\frac{d\theta}{dt} = \frac{d}{dt}\left(\frac{s}{r}\right)$$

Since r is constant, this equation can be rearranged

$$\frac{d\theta}{dt} = \frac{1}{r}\frac{ds}{dt}$$
$$\therefore \omega = \frac{1}{r} \times v$$
$$\therefore v = r\omega$$

(2.7)

. . . .

Equation 2.7 gives us the relationship between the speed of the object (in m s^{-1}) and its angular velocity (in rad s^{-1}). We shall see in the next Topic that this speed is not the same as the linear velocity of the object, since the object is not moving in a straight line, and the velocity describes both the rate and direction at which an object is travelling. (Remember that velocity is a vector quantity.)

Figure 2.3: Tangential speed at several points around a circle

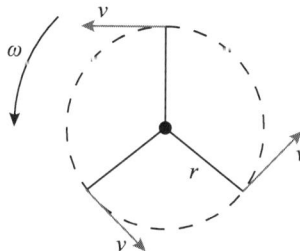

. . . .

Figure 2.3 shows that the speed v at any point is the **tangential speed**, and is always perpendicular to the radius of the circle at that point. If we imagine that Figure 2.3 shows a mass on a string being whirled in a circle, what would happen if the string broke? The mass would continue to travel in a straight line in the direction of the linear speed arrow, that is, it would travel at a tangent to the circle. We will explore this question more fully in the next topic.

Equation 2.7 also shows that there is an important difference between v and ω - that two objects with the same angular velocity can be moving with different tangential speeds. This point is illustrated in the next worked example.

Example

Consider a turntable of radius 0.30 m rotating at constant angular velocity 1.5 rad s^{-1}. Compare the tangential speeds of a point on the circumference of the turntable and a point midway between the centre and the circumference.

The point on the circumference has r = 0.30 m, so

$$v_1 = r_1 \omega$$
$$\therefore v_1 = 0.30 \times 1.5$$
$$\therefore v_1 = 0.45 \text{ ms}^{-1}$$

The point midway between the centre and the circumference is moving with the same angular velocity, but the radius of the motion is only 0.15 m.

$$v_2 = r_2 \omega$$
$$\therefore v_2 = 0.15 \times 1.5$$
$$\therefore v_2 = 0.225 \text{ ms}^{-1}$$

The point on the circumference is moving at twice the speed (in m s^{-1}) of the point with the smaller radius.

. . . .

This difference in tangential speeds is emphasised in Figure 2.4.

Figure 2.4: Points moving at the same angular velocity, but with different tangential velocities

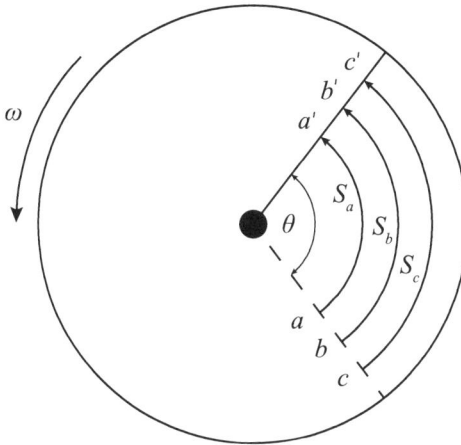

The points a, b and c all lie on the same radius of the circle, which is rotating at angular velocity ω about the centre of the circle. The radius moves through angle θ in time δt. In this same time δt, the points a, b and c move through distances S_a, S_b and S_c respectively, where $S_a < S_b < S_c$. The tangential speeds at each point on the radius are therefore $^{S_a}/_{\delta t}$, $^{S_b}/_{\delta t}$ and $^{S_c}/_{\delta t}$ where $^{S_a}/_{\delta t} < ^{S_b}/_{\delta t} < ^{S_c}/_{\delta t}$.

Compact disc player

There is an online activity showing how the information on a CD is read by the laser in a CD player.

Go online

We can use Equation 2.7 to find an expression relating the angular acceleration to the **tangential acceleration**, the rate at which the speed of an object is changing. Starting with Equation 2.7

$$v = r\omega$$

We can differentiate this equation with respect to time

$$\frac{dv}{dt} = \frac{d}{dt}(r\omega)$$
$$\therefore \frac{dv}{dt} = r\frac{d\omega}{dt}$$
$$\therefore a = r\alpha$$

(2.8)

. . . .

As with the tangential speed, the tangential acceleration depends on the radius of the motion as well as the rate of change of angular velocity. This acceleration is always perpendicular to the radius of the circle.

Quiz: Angular velocity and tangential speed

Go online

Q11: A mass on the end of a 1.20 m length of rope is being swung round in a circle at an angular velocity of 4.00 rad s^{-1}. What is the tangential speed of the mass?

a) 0.300 m s^{-1}
b) 3.33 m s^{-1}
c) 4.80 m s^{-1}
d) 3.33π m s^{-1}
e) 4.80π m s^{-1}

. . . .

Q12: An object is moving in a circular path with angular velocity 12.5 rad s^{-1}. If the speed of the object is 20.0 m s^{-1}, find the radius of the path.

a) 0.255 m
b) 0.625 m
c) 1.60 m
d) 10.1 m
e) 250 m

. . . .

Q13: What is the speed of an object moving in a circle of radius 1.75 m with periodic time 2.40 s?

a) 0.729 m s^{-1}
b) 1.50 m s^{-1}
c) 2.29 m s^{-1}
d) 4.20 m s^{-1}
e) 4.58 m s^{-1}

. . . .

Q14: A spinning disc slows down from ω = 4.50 rad s^{-1} to 1.80 rad s^{-1} in 5.00 s. If the radius of the disc is 0.200 m, find the tangential deceleration of a point on the circumference of the disc.

a) 0.108 m s^{-2}
b) 0.370 m s^{-2}
c) 0.540 m s^{-2}
d) 2.70 m s^{-2}
e) 3.39 m s^{-2}

. . . .

Q15: An object is moving in a circle of radius 0.25 m with a constant angular velocity of 6.4 rad s^{-1}. Through what distance does the object travel in 2.0 s?

a) 1.6 m
b) 3.2 m
c) 10.2 m
d) 12.8 m
e) 51.2 m

. . . .

2.6   Centripetal acceleration

Up until now in our discussion of circular motion, we have concentrated upon the kinematic relationships for circular motion. We will now shift our attention to how Newton's laws of motion apply to circular motion.

We will see that the velocity of an object moving in a circle is constantly changing, hence there is a constant acceleration, requiring a force that acts towards the centre of the circle. This acceleration is quite separate from the tangential acceleration (a) we have met previously.

After discussing Newton's laws we will move on to look at some practical examples of circular motion. We will be paying particular attention to the different forces that are involved, using free-body diagrams to determine the magnitude and direction of these forces.

Consider an object moving in a circle of radius r with constant angular velocity ω. We know that at any point on the circle, the object will have tangential velocity $v = r\omega$. The object moves through an angle $\Delta\theta$ in time Δt.

Figure 2.5 shows the velocity vectors at two points A and B on the circumference of the circle. The tangential velocities v_a and v_b at these points have the same magnitude, but are clearly pointing in different directions. If the angular displacement between A and B is $\Delta\theta$, then the angle between the vectors v_a and v_b is also $\Delta\theta$. Check you can prove this before you proceed.

Figure 2.5: Velocity vectors at two points on a circle

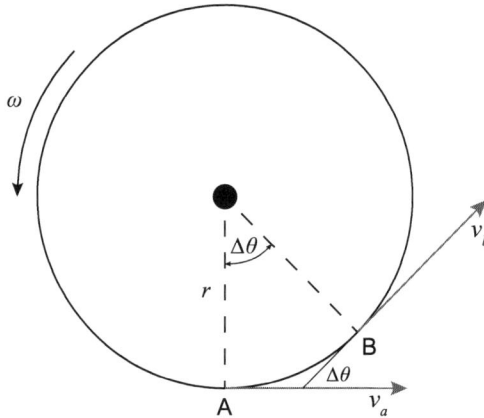

The change in velocity Δv is equal to $v_b - v_a$. We can use a 'nose-to-tail' vector diagram to determine Δ_v, as shown in Figure 2.6. Both v_b and v_a have magnitude v, so the vector XY represents v_b and vector YZ represents $-v_a$.

Figure 2.6: Determination of the change in velocity

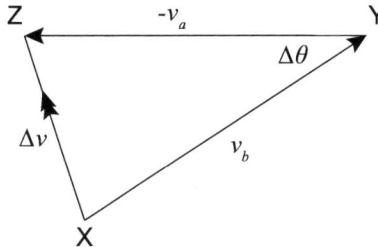

In the limit where Δtt is small, $\Delta\theta$ tends to zero. In this case the angle ZXY tends to 90°, and the vector Δv is **perpendicular** to the velocity, so Δv points towards the centre of the circle. The velocity change, and hence the acceleration, is directed towards the **centre of the circle**. To distinguish this acceleration from any tangential acceleration that may occur, we will denote it by the symbol $a\perp$, since it is perpendicular to the velocity vector.

To calculate the magnitude of $a\perp$, we use the equation

$$a_\perp = \frac{\Delta v}{\Delta t}$$

Since $\Delta\theta$ is small, $\Delta v = v\Delta\theta$, so long as $\Delta\theta$ is measured in radians. The acceleration is therefore

$$a_\perp = \frac{v\Delta\theta}{\Delta t}$$

Taking the limit where Δt approaches zero, this equation becomes

$$a_\perp = \frac{vd\theta}{dt}$$
$$\therefore a_\perp = v\omega$$

Finally we can substitute for $v = r\omega$ to get

$$a_\perp = \frac{v^2}{r} = r\omega^2$$

(2.9)

. . . .

This acceleration $a_\perp$ is called the **centripetal acceleration**. It is always directed towards the centre of the circle, and it must not be confused with the tangential acceleration a we met previously, which occurs when an orbiting object changes its tangential speed. The centripetal acceleration occurs whenever an object is moving in a circular path, even if its tangential speed is constant. (In some textbooks it is referred to as the radial acceleration, as it is always directed along a radius of the circle.)

Examples

1.

Find the centripetal acceleration of an object moving in a circular path of radius 1.20 m with constant tangential speed of 4.00 m s^{-1}.

The centripetal acceleration is calculated using the formula

$$a_\perp = \frac{v^2}{r}$$
$$\therefore a_\perp = \frac{4.00^2}{1.20}$$
$$\therefore a_\perp = 13.3 \text{ m s}^{-2}$$

....

2.

A model aeroplane on a rope 10 m long is circling with angular velocity 1.2 rad s^{-1}. If this speed is increased to 2.0 rad s^{-1} over a 5.0 s period, calculate

1. the angular acceleration;
2. the tangential acceleration;
3. the centripetal acceleration at these two velocities.

1. The data in the question tells us ω_0 = 1.2 rad s^{-1}, ω = 2.0 rad s^{-1}, t = 5.0 s and α is unknown. Using the relationship $\omega = \omega_0 + \alpha t$ we can calculate the angular acceleration α.

$$\omega = \omega_0 + \alpha t$$
$$\therefore \alpha t = \omega - \omega_0$$
$$\therefore \alpha = \frac{\omega - \omega_0}{t}$$
$$\therefore \alpha = \frac{2.0 - 1.2}{5.0}$$
$$\therefore \alpha = 0.16 \text{ rad s}^{-2}$$

2. To calculate the tangential acceleration a, use the relationship between tangential and angular acceleration derived previously.

$$a = r\alpha$$
$$\therefore a = 10 \times 0.16$$
$$\therefore a = 1.6 \text{ m s}^{-2}$$

3. We use the formula $a_\perp = r\omega^2$ to calculate the centripetal acceleration. When $\omega =$ 1.2 rad s^{-1}

$$a_\perp = r\omega^2$$
$$\therefore a_\perp = 10 \times 1.2^2$$
$$\therefore a_\perp = 14.4 \text{ m s}^{-2}$$

When ω = 2.0 rad s^{-1}

$$a_\perp = r\omega^2$$
$$\therefore a_\perp = 10 \times 2.0^2$$
$$\therefore a_\perp = 40 \text{ m s}^{-2}$$

. . . .

Work through the next example to make sure you understand the difference between tangential and centripetal acceleration.

Example

A model aeroplane on a rope 10 m long is circling with angular velocity 1.2 rad s^{-1}. If this speed is increased to 2.0 rad s^{-1} over a 5.0 s period, calculate

1. the angular acceleration;

2. the tangential acceleration;

3. the centripetal acceleration at these two velocities.

1. The data in the question tells us ω_0 = 1.2 rad s^{-1}, ω = 2.0 rad s^{-1}, t = 5.0 s and α is unknown. Using the relationship $\omega = \omega_0 + \alpha t$ we can calculate the angular acceleration α.

$$\omega = \omega_0 + \alpha t$$
$$\therefore \alpha t = \omega - \omega_0$$
$$\therefore \alpha = \frac{\omega - \omega_0}{t}$$
$$\therefore \alpha = \frac{2.0 - 1.2}{5.0}$$
$$\therefore \alpha = 0.16 \text{ rad s}^{-2}$$

2. To calculate the tangential acceleration a, use the relationship between tangential and angular acceleration derived previously.

$$a = r\alpha$$
$$\therefore a = 10 \times 0.16$$
$$\therefore a = 1.6 \text{ m s}^{-2}$$

3. We use the formula $a_\perp = r\omega^2$ to calculate the centripetal acceleration. When $\omega =$ 1.2 rad s^{-1}

$$a_\perp = r\omega^2$$
$$\therefore a_\perp = 10 \times 1.2^2$$
$$\therefore a_\perp = 14.4 \text{ m s}^{-2}$$

When ω = 2.0 rad s^{-1}

$$a_\perp = r\omega^2$$
$$\therefore a_\perp = 10 \times 2.0^2$$
$$\therefore a_\perp = 40 \text{ m s}^{-2}$$

....

Quiz: Centripetal acceleration

Go online

Q16: A disc of radius 0.50 m is rotating with angular velocity 4.1 rad s^{-1}. What is the centripetal acceleration of a point on its circumference?

a) 2.1 m s^{-2}
b) 4.2 m s^{-2}
c) 8.4 m s^{-2}
d) 17 m s^{-2}
e) 34 m s^{-2}

....

Q17: Which *one* of the following statements is true?

a) The centripetal acceleration is always at a tangent to the circle.
b) The centripetal acceleration is always directed towards the centre of the circle.
c) There is only a centripetal acceleration when the tangential speed is changing.
d) The centripetal acceleration does not depend on the angular velocity.
e) The centripetal acceleration does not depend on the radius of the circular motion.

....

Q18: A car takes a corner at 15 m s^{-1}. If the radius of the corner is 40 m, what is the centripetal acceleration of the car?

a) 0.18 m s^{-2}
b) 5.6 m s^{-2}
c) 107 m s^{-2}
d) 600 m s^{-2}
e) 9000 m s^{-2}

....

Q19: A particle moves in a circular path of radius 0.10 m at a constant rate of 6.0 revolutions per second. What is the centripetal acceleration of the particle?

a) 3.6 m s^{-2}
b) 3.8 m s^{-2}
c) 14 m s^{-2}
d) 36 m s^{-2}
e) 140 m s^{-2}

. . . .

Q20: An object is moving in a circular path with speed v m s^{-1}. If the radius of the path doubles whilst the speed v remains constant, what happens to the centripetal acceleration?

a) The centripetal acceleration halves in value.
b) The centripetal acceleration doubles in value.
c) The new centripetal acceleration equals the square of the original value.
d) The new centripetal acceleration equals the square root of the original value.
e) The centripetal acceleration remains constant.

. . . .

2.7 Centripetal force

Newton's second law of motion tells us that if an object is undergoing acceleration, then a net force must be acting on the object in the direction of the acceleration. Since we have a centripetal acceleration acting towards the centre of the circle, there must be a **centripetal force** acting in that direction.

Newton's law can be summed up by the equation

$$F = ma$$

If the centripetal acceleration is given by $a_{\perp} = {v^2}/{r} = r\omega^2$, then the centripetal force acting on a body of mass m moving in a circle of radius r is

$$F = \frac{mv^2}{r} = mr\omega^2$$

(2.10)

. . . .

Figure 2.7: Centripetal force

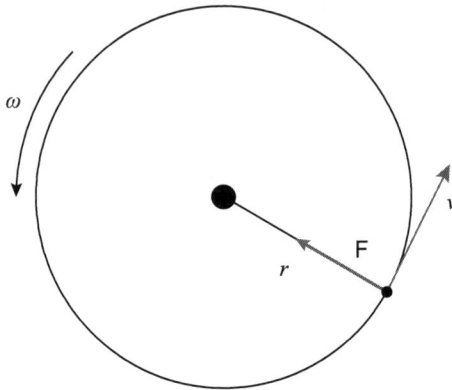

....

This is the force which must act on a body to make it move in a circular path. If this force is suddenly removed, the body will move in a straight line at a tangent to the circle with speed v, since there will be no force acting to change the velocity of the body.

Example

Compare the centripetal forces required for a 2.0 kg mass moving in a circle of radius 40 cm if the velocity is:

1. 3.0 m s⁻¹;

2. 6.0 m s⁻¹.

1. Using Equation 2.10, the force required is

$$F = \frac{mv^2}{r}$$
$$\therefore F = \frac{2.0 \times 3.0^2}{0.40}$$
$$\therefore F = 45 \text{ N}$$

2. In this case

$$F = \frac{mv^2}{r}$$
$$\therefore F = \frac{2.0 \times 6.0^2}{0.40}$$
$$\therefore F = 180 \text{ N}$$

So doubling the velocity means that the centripetal force required increases by a factor of four, since $F \propto v^2$.

. . . .

2.7.1   Object moving in a horizontal circle

Consider the case shown in Figure 2.8(a) of an object moving at constant angular velocity ω in a horizontal circle, such as a mass being whirled overhead on a string.

Figure 2.8: (a) An object moving in a horizontal circle; (b) horizontal force acting on the object

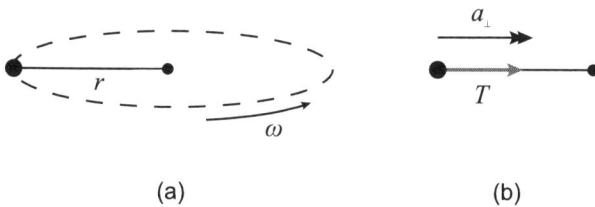

(a)   (b)

. . . .

In the horizontal direction, Figure 2.8(b) shows there is an acceleration $a_\perp$ acting towards the centre of the circle. The only force acting in the horizontal direction is the tension in the string, so this tension must provide the centripetal force $mr\omega^2$. So in this case

$$T = mr\omega^2$$

(2.11)

. . . .

2.7.2   Vertical motion

Let us now consider the same object being whirled in a vertical circle. Figure 2.9 shows the object at three points on the circle, with the forces acting at each point.

Figure 2.9: Object undergoing circular motion in a vertical plane

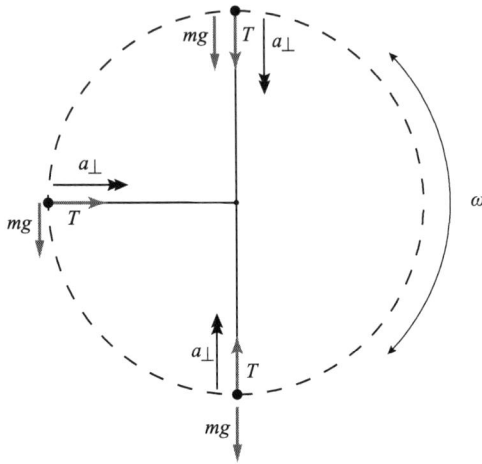

If we consider the forces acting when the object is at the top of the circle, we find that both the weight and the tension in the string are acting downwards. Thus the resultant force acting towards the centre of the circle is

$$T_{top} + mg = mr\omega^2$$

This resultant force must provide the centripetal force to keep the object moving in its circular path. Hence

$$T_{top} = mr\omega^2 - mg$$

(2.12)

On the other hand, when the object reaches the lowest point on the circle, the tension and the weight are acting in opposite directions, so the resultant force acting towards the centre of the circle is

$$T_{bottom} - mg = mr\omega^2$$

Again, this force supplies the centripetal force so in this case

$$T_{bottom} = mr\omega^2 + mg$$

(2.13)

. . . .

Equation 2.12 and Equation 2.13 give us the minimum and maximum values for the tension in the string.

When the string is horizontal, there is no component of the weight acting towards the centre of the circle, so the tension in the string provides all the centripetal force,

$$T_{horiz.} = mr\omega^2$$

(2.14)

. . . .

Motion in a vertical circle

A man has tied a 1.2 kg mass to a piece of rope 0.80 m long, which he is twirling round in a vertical circle, so that the rope remains taut. He then starts to slow down the speed of the mass.

1. At what point of the circle is the rope likely to go slack?

2. What is the speed (in m s^{-1}) at which the rope goes slack?

. . . .

Quiz: Horizontal and vertical motion

Go online

Q21: What is the centripetal force acting on a 2.5 kg mass moving in a circle of radius 4.0 m when its speed is 8.0 m s^{-1}?

a) 5.0 N
b) 6.3 N
c) 40 N
d) 102 N
e) 640 N

. . . .

Q22: A centripetal force of 25 N causes a 1.4 kg mass to move in a circular path of radius 0.50 m. What is the angular velocity of the mass?

a) 1.8 rad s^{-1}
b) 3.0 rad s^{-1}
c) 6.0 rad s^{-1}
d) 8.9 rad s^{-1}
e) 36 rad s^{-1}

. . . .

Q23: Consider an object of fixed mass m moving in a circle with a constant tangential speed v. Which *one* of the following statements is true?

a) Since the tangential speed is constant, there is no centripetal force.
b) The centripetal force increases if the radius increases.
c) The centripetal acceleration increases if the centripetal force decreases.
d) If the radius of the circle is increased, the centripetal force decreases.
e) There is no centripetal acceleration.

. . . .

Q24: A 1.20 kg mass on a 2.00 m length of string is being whirled in a horizontal circle. If the maximum tension in the string is 125 N, what is the maximum possible speed of the mass?

a) 8.66 m s^{-1}
b) 14.4 m s^{-1}
c) 17.3 m s^{-1}
d) 20.4 m s^{-1}
e) 208 m s^{-1}

. . . .

Q25: A mass on a string is being whirled around in a vertical circle. At which point in its motion is the string most likely to snap?

a) When the mass is at the bottom of the circle.
b) When the mass is at the top of the circle.
c) When the string is horizontal.
d) When the string is at 45° to the horizontal.
e) There is an equal chance of the string snapping everywhere around the circle.

. . . .

2.8 Applications

The centripetal acceleration and centripetal force of an object moving in a circular path depend on the speed or angular velocity of the object and the radius of the circle. Without this force, the object would move off in a straight line at a tangent to the circle.

We will now calculate the centripetal force and acceleration in some practical applications. In each case a free-body diagram will be used to calculate the tensions and other forces that supply the centripetal force. Before carrying out the self-assessment test for this topic you should make sure you understand the forces involved in each of the examples. Try to think of other situations involving circular motion, and the forces that provide the centripetal force.

2.8.1 Rollercoaster

When a person rides on a rollercoaster that follows a loop the loop track, they often feel very "light" at the top of the loop. However, their weight does not change throughout the motion. In fact, it is the normal reaction force applied to them by the seat which alters. It is this which causes the strange sensation.

Assume the rollercoaster is moving at a constant speed, then the centripetal force required to keep them moving in a circle will also be constant. However, when they are at the top of the loop, the centripetal force is provided by both their weight and the normal reaction force. So the normal reaction force is very small and the person feels "light".

Rollercoaster

There is an animation online displaying the forces acting on a person throughout a rollercoaster ride.

Go online

At the bottom

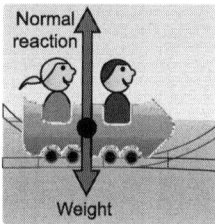

$$\frac{mv^2}{r} = Normal\ reaction\ -\ Weight$$

$$Normal\ reaction\ =\ \frac{mv^2}{r} + Weight$$

At the side

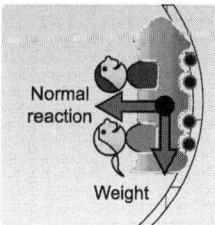

$$\frac{mv^2}{r} = Normal\ reaction$$

At the top

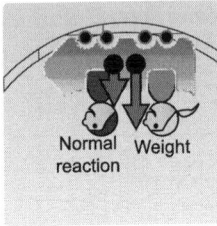

$$\frac{mv^2}{r} = Normal\ reaction + Weight$$

$$Normal\ reaction = \frac{mv^2}{r} - Weight$$

. . . .

Infact, if the speed is sufficiently large, then the normal reaction force applied to the person at the top of the loop will become zero and the person will feel weightless. At this point

$$\frac{mv^2}{r} = mg$$

Simplifying and rearranging, this becomes

$$v^2 = gr$$

Astronauts train in aircrafts that follow a circular path like this. At the top of a circular manoeuvre, they feel weightless. This allows them to become accustomed to experiencing the sensation of zero gravity. Since this was found to have an unsettling effect on their stomachs, such aircrafts were nicknamed "vomit comets".

2.8.2 Conical pendulum

The next situation we will study is the **conical pendulum** - a pendulum of length l whose bob moves in a circle of radius r at a constant height. Figure 2.10 shows such a pendulum, with a free-body diagram of all the forces acting on the bob.

Figure 2.10: (a) Conical pendulum moving in a horizontal circle (b) free-body diagram showing the forces acting on the bob

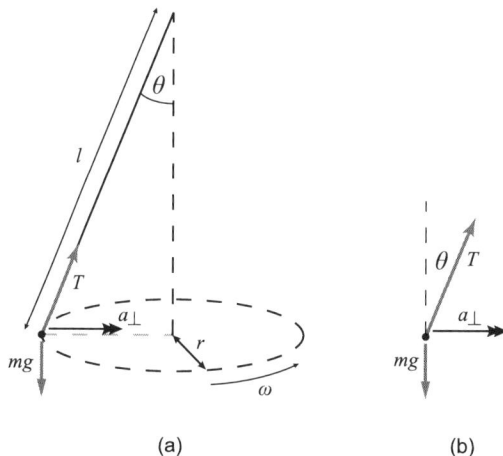

(a)   (b)

. . . .

The string of the pendulum makes an angle θ with the vertical, such that $\sin \theta = {r}/{l}$. To calculate this angle, we use the free-body diagram in Figure 2.10(b) and resolve vertically and horizontally.

Vertically, the system is in equilibrium, so

$$T \cos \theta = mg$$

Horizontally, we have an acceleration towards the centre of the circle, so there must be a centripetal force which is provided by the horizontal component of the tension, $T \sin \theta$, so

$$T \sin \theta = mr\omega^2$$
$$\therefore T \sin \theta = ml \sin \theta \omega^2$$
$$\therefore T = ml\omega^2$$

We can substitute for $T = ml\omega^2$ in the first equation, which gives us

$$ml\omega^2 \cos\theta = mg$$
$$\therefore l\omega^2 \cos\theta = g$$
$$\therefore \cos\theta = \frac{g}{l\omega^2}$$

(2.15)

. . . .

We have an inverse square dependence of $\cos\theta$ on the angular velocity. If ω increases, $\cos\theta$ decreases and hence θ increases. The faster the bob is moving, the closer to horizontal the string becomes.

Example

Consider a conical pendulum of length 1.0 m. Compare the angle the string makes with the vertical when the pendulum completes exactly 1 revolution and 2 revolutions per second ($\omega = 2\pi$ rad s^{-1} and $\omega = 4\pi$ rad s^{-1}).

In the first instance, when $\omega = 2\pi$ rad s^{-1}

$$\cos\theta = \frac{g}{l\omega^2}$$
$$\therefore \cos\theta = \frac{9.8}{1.0 \times (2\pi)^2}$$
$$\therefore \cos\theta = \frac{9.8}{4\pi^2}$$
$$\therefore \cos\theta = 0.248$$
$$\therefore \theta = 76°$$

When $\omega = 4\pi$ rad s^{-1}

$$\cos\theta = \frac{g}{l\omega^2}$$
$$\therefore \cos\theta = \frac{9.8}{1.0 \times (4\pi)^2}$$
$$\therefore \cos\theta = \frac{9.8}{16\pi^2}$$
$$\therefore \cos\theta = 0.062$$
$$\therefore \theta = 86°$$

Clearly the angle θ approaches $90°$ as the speed of the pendulum bob gets higher and higher.

. . . .

Conical pendulum

This simulation shows a practical example of a conical pendulum.

. . . .

Go online

2.8.3   Cars cornering

When a car takes a corner, the frictional force between the car wheels and the road provides the centripetal force. If there is insufficient friction the car will skid. If we know the size of the frictional force, we can calculate the maximum speed at which a car can safely negotiate a corner.

Figure 2.11: (a) A car taking a corner; (b) the forces acting on the car

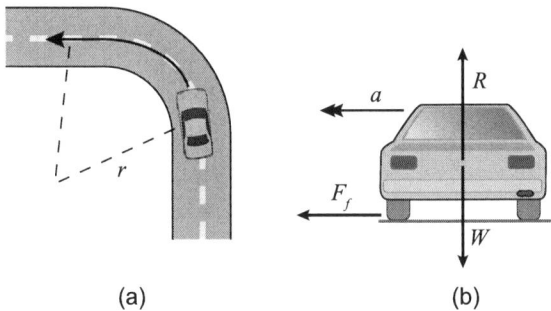

(a)   (b)

. . . .

Figure 2.11(a) shows a car of mass m taking a corner. The radius of the bend is r. A free-body diagram of the forces acting on the car is shown in Figure 2.11(b). The centripetal force is provided by the frictional force F_f. Resolving vertically, using Newton's third law of motion, the normal reaction force R is equal to the weight $W = mg$. The frictional force provides the centripetal force acting on the car. If $F_{f\,\text{max}}$ is the maximum value of the frictional force, then the maximum speed at which the car can take the corner is given by

$$F_{f\,\text{max}} = \frac{m v_{\text{max}}^2}{r}$$

(2.16)

. . . .

It is worth noting that the maximum value of the frictional force becomes extremely small when the road is icy, therefore reducing the maximum speed at which the car can take the corner.

2.8.4   Banked tracks

We saw that the safe cornering of a car on a flat road depends on whether the tyres and the road surface can provide sufficient friction. However, if a vehicle is on a banked track, then it can in fact still successfully take a corner when there is no frictional force present. Banked tracks are sloped at an angle. This means the horizontal component of the normal reaction force provides the required centripetal force. The vertical component of the normal reaction force balances the weight.

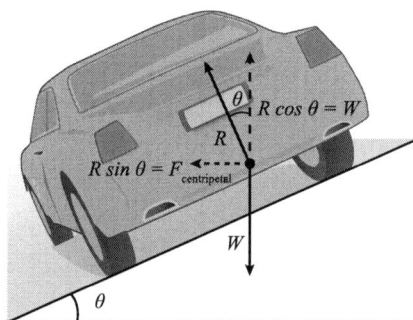

So for a track sloped at an angle θ to the horizontal

$$R\sin\theta = \frac{mv^2}{r}$$
$$R\cos\theta = mg$$

Dividing the top equation by the bottom gives

$$\tan\theta = \frac{v^2}{gr}$$

So

$$v = \sqrt{gr\tan\theta}$$

Looking at the equation above, we can see that a steeper bank allows the corner to be taken at a greater speed.

Banked corners

There is an animation online looking at a bobsleigh taking a banked corner on ice.

....

Go online

A velodrome is a bicycle racing track that consists of two 180° circular bends connected by two straight sections. It is another example of a banked track.

2.8.5 Funfair rides

A popular amusement park ride is called the sticky wall or the rotor. This involves a drum that spins around a vertical axis. People stand with their backs to the inside wall and the drum spins. Once it is rotating with a sufficiently large angular velocity, the floor drops away. Everyone has the sensation that they are "stuck" or "pinned" to the wall, but what is really happening is the drum wall is exerting a normal reaction force. This provides the centripetal force to keep them moving in a circular path. They do not fall down since the friction acting upwards balances their weight.

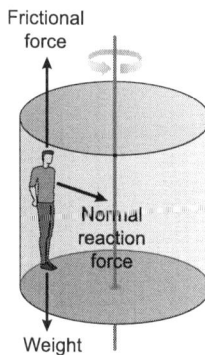

Frictional force

Normal reaction force

Weight

Quiz: Conical pendulum and cornering

Go online

Q26: The string of a conical pendulum makes an angle θ with the vertical, and the tension in its string is T. What is the centripetal force acting on the bob, in terms of T and θ?

a) $T \times \theta$

b) $T \times \cos\theta$

c)
$$T/_{\cos\theta}$$

d) $T \times \sin\theta$

e)
$$T/_{\sin\theta}$$

. . . .

Q27: Consider a conical pendulum of length 50 cm. What angle does the string make with the vertical when the angular speed of the pendulum is 6.0 rad s^{-1}?

a) 0.55°

b) 33°

c) 57°

d) 87°

e) 89°

. . . .

Q28: The conical pendulum in question 11 is rotating at constant angular velocity. What is the period of the pendulum?

a) $2\pi \frac{g}{l\cos\theta}$

b) $2\pi \sqrt{g/_{l\cos\theta}}$

c) $\sqrt{l\cos\theta/_{g}}$

d) $2\pi \frac{l\cos\theta}{g}$

e) $2\pi \sqrt{l\cos\theta/_{g}}$

. . . .

Q29: For a car cornering on a flat (unbanked) road, the

a) centripetal force is provided by the frictional force.

b) centripetal force is provided by the normal reaction force.

c) centripetal force is provided by the weight.

d) normal reaction force is equal to the frictional force.

e) weight is equal to the frictional force.

. . . .

Q30: A car can take a corner on a banked road at a higher speed than on a horizontal road because

a) the weight of the car increases.
b) less centripetal force is required.
c) the radius of the corner decreases.
d) a component of the normal reaction force contributes to the centripetal force.
e) the mass of the car decreases.

. . . .

2.9 Extended information

Web links

There are web links available online exploring the subject further.

. . . .

2.10 Summary

─┤ Summary ├─

You should now be able to:

- measure angles and angular displacement in radians, and convert an angle measured in degrees into radians, and vice versa;

- use angular displacement, angular velocity and angular acceleration to describe motion in a circle;

- apply the equation $T = \frac{2\pi}{\omega}$ relating the periodic time to the angular velocity;

- derive the angular kinematic relationships, and apply them to solving problems of circular motion with uniform angular acceleration;

- relate the angular velocity to the tangential (linear) speed of a body moving in a circle, and derive the equation $v = r\omega$;

- apply the expression $a = r\alpha$ relating tangential and angular accelerations;

- derive the expressions for centripetal motion in terms of v and ω;

- calculate the centripetal acceleration and centripetal force of an object undergoing circular motion;

- use free-body diagrams to calculate centripetal forces;

- describe a number of real-life situations in which the centripetal force plays an important role.

2.11 Assessment

End of topic 2 test

The following test contains questions covering the work from this topic.

Go online

Q31: How many radians are equivalent to 66°?

....

Q32: A flywheel is rotating with constant angular velocity 21 rad s^{-1}.

Calculate the time taken for the flywheel to complete exactly 6 revolutions.

Time taken = _____ s

....

Q33: A moon orbiting a distant planet has a period of 24 days.

Calculate the angular velocity of the planet, in rad s^{-1}.

Angular velocity = _____ rad s^{-1}

....

Q34: An electric fan is switched from a "Low" setting (angular velocity 4.9 rad s^{-1}) to a "High" setting (angular velocity 15 rad s^{-1}).

The angular acceleration is 6.6 rad s^{-2}.

Calculate the time taken for the fan to reach the higher angular velocity.

t = _____ s

....

Q35: As the brakes are applied to a bicycle wheel, the angular velocity of the wheel changes from 23.50 rad s^{-1} to 5.00 rad s^{-1}.

The wheel turns through exactly 10 revolutions whilst the brakes are being applied.

Calculate the magnitude of the angular deceleration of the wheel.

Angular deceleration = _____ rad s^{-2}

....

Q36: Passengers in a Ferris Wheel sit in cars 11 m from the centre of the wheel. The wheel takes 100 s to complete one revolution, travelling with uniform angular velocity.

Calculate the speed of a passenger car.

Speed = _____ m s^{-1}

....

Q37: A cyclist travelling around a circular track of radius 32 m increases his speed from 13 m s^{-1} to 25 m s^{-1} in 5.0 s.

Calculate the average angular acceleration of the cyclist.

Average angular acceleration = _____ rad s^{-2}

....

Q38: A cyclist is riding in a circle of radius 17.0 m at an average speed of 9.5 m s^{-1}.
Calculate the cyclist's centripetal acceleration.
Centripetal acceleration = _____ m s^{-2}

....

Q39: A disc is rotating at constant angular velocity about its centre. At the edge of the disc, the centripetal acceleration is 6.9 m s^{-2}, whilst at a point 6.5 cm from the centre of the disc, the centripetal acceleration is 5.2 m s^{-2}.
Calculate the radius of the disc.
Radius = _____ cm

....

Q40: A 1.25 kg mass is moving in a circular path of radius 1.82 m, with periodic time 1.95 s.
Calculate the centripetal force acting on the mass.
Centripetal force = _____ N

....

Q41: A model aeroplane (mass 0.80 kg) can fly at up to 28 m s^{-1} in a horizontal circle on a guideline which has a breaking tension of 100 N.
Calculate the minimum radius in which the plane could be flown if it is travelling at its maximum speed.
Minimum radius = _____ m

....

Q42: A conical pendulum consists of a string of length 1.1 m and a bob of mass 0.34 kg. The string makes an angle of 25° with the vertical.
Calculate the tension in the string and the angular velocity of the pendulum.

1. Tension = _____ N
2. Angular velocity = _____ rad s^{-1}

....

Q43: A motorcyclist is approaching a hump-backed bridge, as shown in the diagram. The bridge forms part of a circle of radius r = 20.0m. The combined mass of the motorcycle and rider is 165 kg.

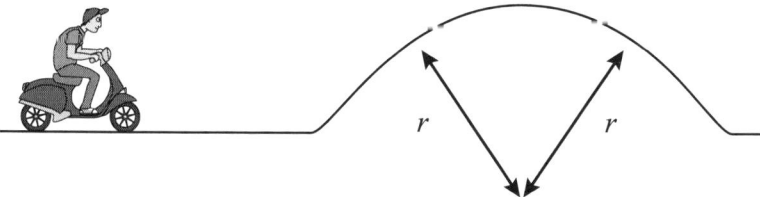

Calculate the maximum speed at which the rider could travel without leaving the road at the top of the bridge.

Maximum speed = _____ m s^{-1}

. . . .

Q44: The static frictional force between a car of mass 1990 kg and a dry road has a maximum value of 1.71×10^4 N.

Calculate the maximum speed at which the car could take an unbanked bend of radius 40.0 m without skidding.

Maximum speed = _____ m s^{-1}

. . . .

Topic 3

Rotational dynamics

Contents

Prerequisite knowledge

- *Applying Newton's laws of motion.*
- *Angular velocity and acceleration (Unit 1 - Topic 2).*
- *Understanding of the principle of conservation of linear momentum.*

Learning objectives

By the end of this topic you should be able to:

- *state that the moment (torque) of a force is the tendency of a force to cause or change rotational motion of an object;*
- *state and apply the equation $T = Fr$,*
- *state that an unbalanced torque acting on an object produces an angular acceleration;*
- *state that the moment of inertia of an object is a measure of its resistance to angular acceleration about a given axis;*
- *explain that the moment of inertia of an object about an axis depends on the mass of the object, and the distribution of the mass about the axis;*

- *state that the moment of inertia I of a point mass of mass m at a distance r from a fixed axis is given by the equation $I = mr^2$;*

- *use the relevant equation to calculate the moment of inertia for the following shapes- hoop about centre, rod about centre, rod about end, disc about centre, sphere about centre;*

- *calculate the moment of inertia of a combination of rotating bodies;*

- *state and apply the equation $T = I\alpha$.*

3.1   Introduction

In the previous topic we derived kinematic relationships for describing circular motion which compared directly to the linear kinematic relationships. In this topic, we will now perform a similar exercise for other aspects of rotational dynamics. We will begin by looking at the turning effect (torque) of a force being used to make an object rotate about an axis. This will lead us to the concept of moment of inertia.

3.2   Torque and moment

We will begin by studying the turning effect of a force, and the example we shall look at is a force being used to push a door open. The axis of rotation is the door hinge. What happens if you push the door in the same place with different forces?

Figure 3.1: (a) Small force, and (b) large force used to push open a door

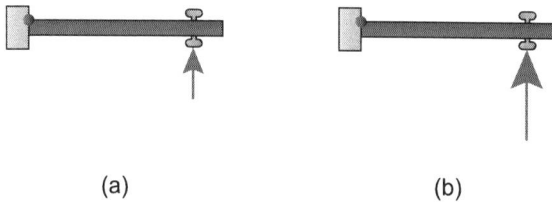

(a)   (b)

Since $F = ma$, the door opens more quickly the larger the force you use, something you can easily check for yourself! The turning effect increases when the force is increased.

Now think about what happens if you apply the same force near the hinge or near the edge of the door.

Figure 3.2: Same force being applied (a) far from the axis, and (b) close to the axis

(a)   (b)

The force has a greater turning effect when it is applied near the edge of the door - it is harder to open the door when you push it near the hinge. So the turning effect also depends on how far from the axis of rotation the force is being applied.

The turning effect of a force is called the **torque** T or **moment**, and is defined by

$$T = Fr$$

(3.1)

. . . .

In this equation F is the size of the force and r is the distance between the axis and the point where the force is being applied. Since the torque is producing a circular motion, r is the radius of the motion. T has the units N m. Looking at Figure 3.3 below, we can see that Equation 3.1 does not quite tell the full story. This equation is only valid when the force is applied at right angles to the radius.

Figure 3.3: Same force applied (a) perpendicular to the door, and (b) at an angle to the door

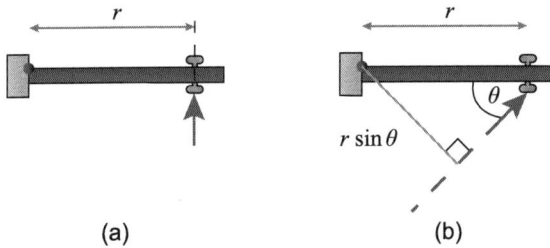

(a)   (b)

. . . .

The dashed line in Figure 3.3 represents the **line of action** of the force. When calculating the torque, the important distance is not where the force is applied, but the **perpendicular distance from the line of action to the axis of rotation**. If F makes an angle θ to the line r joining the axis to the point where the force is applied, then the torque is given by

$$T = Fr\sin\theta$$

(3.2)

. . . .

Example

A gardener picks up her wheelbarrow to wheel it along by applying a perpendicular force of 50 N to the handles. If the handle grips are 1.2 m from the axis of the wheel, what is the torque that she applies?

The situation is shown schematically in Figure 3.4

Figure 3.4: Torque applied to the wheelbarrow

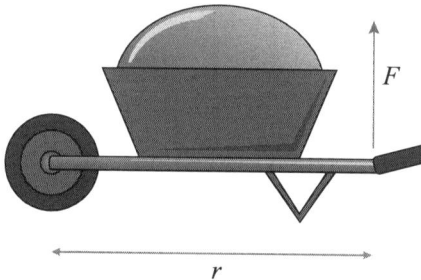

Since a perpendicular force is applied, we can use Equation 3.1

$$T = Fr$$
$$\therefore T = 50 \times 1.2$$
$$\therefore T = 60\,\mathrm{N\,m}$$

. . . .

Quiz: Torques

Q1: Which of the following sets of dimensions could be used to measure a torque?

a) kg m
b) kg m²
c) N kg m²
d) N m⁻¹
e) N m

Go online

. . . .

Q2: What is the torque when a 35 N force is applied at a distance 1.4 m from an axis, perpendicular to the radius?

a) 0.04 N m
b) 25 N m
c) 35 N m
d) 49 N m
e) 69 N m

....

Q3: A 16.0 N force produces a 60.0 N m torque when applied at right angles to the radius of rotation. How far from the axis is the force applied?

a) 3.75 m
b) 26.7 m
c) 48.0 m
d) 60.0 m
e) 960 m

....

Q4: A force of 12.0 N is applied 0.750 m from the axis to produce a torque. The force is applied at an angle of $70°$ to the radius of motion. What is the size of the torque?

a) 0.156 N m
b) 3.08 N m
c) 8.46 N m
d) 9.00 N m
e) 630 N m

....

Q5: A torque of 48.0 N m is produced when a force is applied 0.500 m from an axis. If the force makes an angle of $60°$ with the radius, what is the magnitude of the force?

a) 20.8 N
b) 83.1 N
c) 96.0 N
d) 111 N
e) 192 N

....

3.3   Newton's laws applied to rotational dynamics

We are used to the idea of forces in equilibrium producing no acceleration, and a resultant force producing an acceleration in the direction of the force. A similar situation exists when we are applying torques instead of forces. A system of torques in equilibrium will produce no net turning effect, whilst a resultant torque produces an angular acceleration about the axis.

3.3.1 Torques in equilibrium

To consider the equilibrium situation, let us return to the problem of the wheelbarrow. We will assume the barrow stays horizontal, and we will remove its support so that the gardener has to apply a perpendicular force to keep the barrow horizontal. Suppose the barrow is loaded with 20 kg of rubble, and the centre of mass of this rubble is 50 cm from the wheel axis. If we can neglect the mass of the barrow, what force is necessary to keep it horizontal?

Figure 3.5: Torques in equilibrium

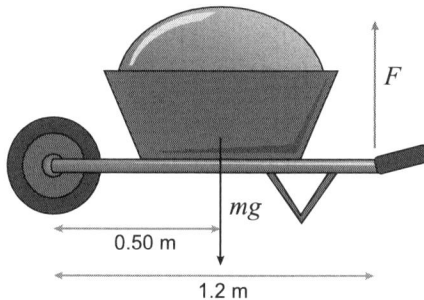

For the above system to be in equilibrium, the clockwise torque (due to the mass of the rubble) must equal the anti-clockwise torque (due to the gardener applying a force). Therefore

$$T_{a-c} = T_c$$
$$\therefore F \times 1.2 = mg \times 0.50$$
$$\therefore F \times 1.2 = 20 \times 9.8 \times 0.50$$
$$\therefore F \times 1.2 = 98$$
$$\therefore F = 82\,\text{N}$$

The problem is similar to a Newton's first law problem, and the method of solving it is the same: draw a free-body diagram to show all the forces acting. The difference is that we are no longer considering point objects, so we must also include where the forces are acting on the diagram. In the above problem, we have been told to ignore the weight of the barrow. In a situation where this cannot be ignored, the weight acts through the centre-of-mass of the object, and the distance from the axis to the centre-of-mass must be known to calculate the torque. The next example illustrates how to solve more complicated problems.

Example

A serving table of mass 2.5 kg and length 80 cm is hinged to a wall, and is supported by a chain which makes an angle of 50° with the horizontal table top. The chain is attached to the edge of the table furthest from the wall (r_c = 80 cm). The table is uniform, so its centre-of-mass is r_t = 40 cm from the wall. A full serving dish of mass 1.0 kg is placed on the table r_d = 60 cm from the wall. Calculate the tension in the chain.

Figure 3.6: Free-body diagram showing the forces acting on the table

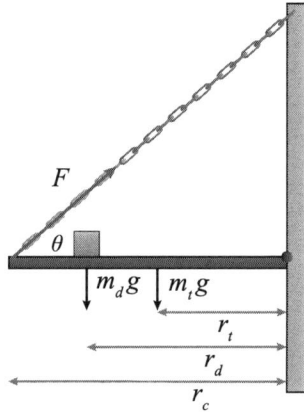

In equilibrium, the anti-clockwise torques must balance the clockwise torques. This means the torques due to the two masses are balanced by the torque due to the tension in the chain.

$$F \times \sin \theta \times r_c = (m_d \times g \times r_d) + (m_t \times g \times r_t)$$
$$\therefore F \times 0.766 \times 0.80 = (1.0 \times 9.8 \times 0.60) + (2.5 \times 9.8 \times 0.40)$$
$$\therefore 0.6128F = 5.88 + 9.8$$
$$\therefore 0.6128F = 15.68$$
$$\therefore F = 25.6\,N$$

If you try this calculation again with the dish at a different position, you will find that the tension in the chain depends on whereabouts on the table the dish is positioned. If the dish is placed on the edge of the table, (r_d = 80 cm), then the tension F = 28.8 N, whereas placing the dish at r_d = 20 cm from the wall reduces F to 19.2 N. Clearly, when we are considering torques, we are not just interested in the mass of an object or objects, but how that mass is distributed relative to the axis.

Torque and static equilibrium

In questions 1 to 4, try to balance the beam by making sure that the moment (sum of force x distance) is the same on each side of the fulcrum (pivot). The length of the beam is 8.0 m and the fulcrum is at the centre of the beam.

Go online

Q6:

Load A = 450N, sits 4 m from the fulcrum.

Where should B = 600N sit?

To the left or right?

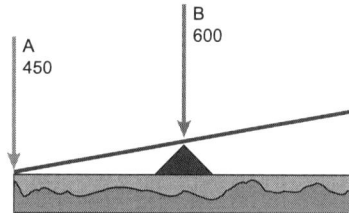

....

Q7:

Load A now moves to 3 m from the fulcrum.

Load B distance = _____ ?

To the left or right?

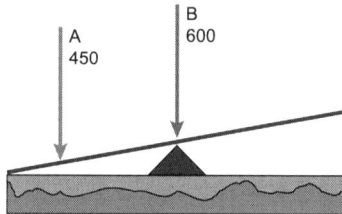

....

Q8:

Load C = 480N is added on A's side, sitting 2 m from the fulcrum.

Load B distance = _____ ?

To the left or right?

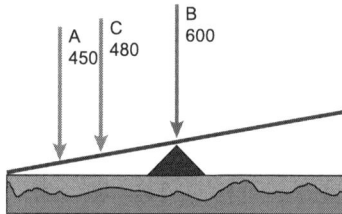

....

Q9:

Load D = 360N is added on B's side, 4 m from fulcrum.

Load B distance = _____ ?

To the left or right?

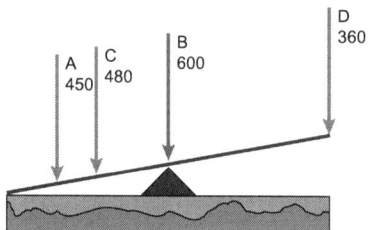

. . . .

3.3.2 Angular acceleration and moment of inertia

Let us now consider a Newton's second law-type problem. We are used to a force producing an acceleration, and the relationship $F = ma$. We will now see if there is an equivalent expression involving torques and angular acceleration. The problem we will look at is that of an object moving in a horizontal circle of fixed radius r at an angular velocity ω.

Figure 3.7: (a) Object moving around a circle; (b) a tangential force applied to the object

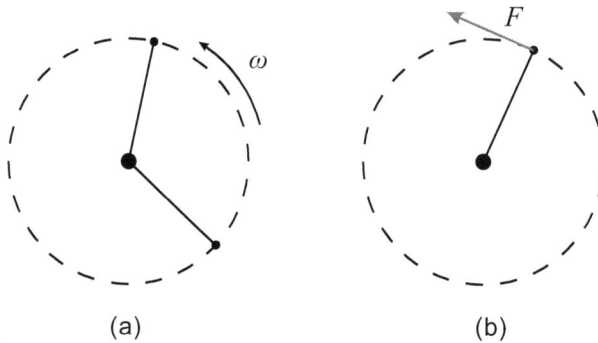

(a)   (b)

. . . .

If we apply an anticlockwise force F to the object at a tangent to the circle, as shown in Figure 3.7(b), the object will accelerate. We can use Newton's second law

$$F = ma$$

We can express this acceleration as an angular acceleration using the expression $a = r\alpha$. Hence

$$F = mr\alpha$$

Since the force is being applied at a distance r from the axis of rotation, perpendicular to the radius, we can substitute for F using $T = Fr$

$$\frac{T}{r} = mr\alpha$$
$$\therefore T = mr^2\alpha$$
$$\therefore T = I\alpha$$

(3.3)

. . . .

We have used the symbol I to represent mr^2 to give us an expression in Equation 3.3 which looks similar to $F = ma$. The quantity I is called the **moment of inertia**, and for a point mass m rotating with radius r about an axis.

$$I = mr^2$$

(3.4)

. . . .

I has the units kg m^2. Throughout this topic we will be using I in our calculations, as it takes into account not only the mass of an object, but also how far the mass is from the axis of rotation.

There are some important properties about the moment of inertia that you must make sure you understand. First, we cannot state a fixed value of I for any object. The value of I depends not only on the mass of the object, but the position of the axis. The axis **must** be specified when giving the moment of inertia of an object. Although there are similarities between mass and moment of inertia, there is a major difference in that the mass of a body is constant but the moment of inertia changes for different axes of rotation.

Secondly, the equation $I = mr^2$ is only valid for a point mass with a known position. For a collection of masses m_1, m_2, $m_3...m_i$, the total moment of inertia is

$$I_{\text{total}} = m_1r_1{}^2 + m_2r_2{}^2 + m_3r_3{}^2 + ... = \sum_i m_i r_i{}^2$$

(3.5)

. . . .

3.4   Calculating the moment of inertia

You should be aware that a satellite can be considered as a point mass, since it is so small compared to the radius of its orbit. This means that the moment of inertia of a satellite can be found from the equation $I = mr^2$.

It is also worth noting that the mass of a thin hoop is all located at the same distance r from its central axis. So, a thin hoop's moment of inertia can also be found using the equation $I = mr^2$.

Finding the moment of inertia for other rotating bodies is considerably more complicated. Using a calculus approach, the body is divided into infinitely small sections. I is calculated for each section and then an integration is carried out to find the total moment of inertia. Such a calculation is beyond the scope of this course. However, you will be expected to employ the following equations for a number of standard shapes. You will need to remember to use the same formula for a hoop as for a point mass, but the other equations should be provided in the exam.

Table 3.1: The moment of inertia

Shape	Axes of rotation	Equation for moment of inertia
Hoop about centre		$I = mr^2$
Rod about centre		$I = \frac{1}{12}ml^2$
Rod about end		$I = \frac{1}{3}ml^2$
Disc about centre		$I = \frac{1}{2}mr^2$

Shape	Axes of rotation	Equation for moment of inertia
Sphere about centre		$I = \frac{2}{5}mr^2$

Moment of inertia

There is animated version available online showing different axes of rotation.

Go online

Use the relevant information from Table 3.1 for following examples.

Examples

1.

A metre stick of length 80 cm and mass 120 g is rotating about one end.
Calculate its moment of inertia.

$$I = \frac{1}{3}ml^2$$
$$I = \frac{1}{3} \times 0.12 \times 0.8^2$$
$$I = 0.0256 \text{ kg m}^2$$

2.

A baton of length 80 cm and mass 120 g is rotating about its centre.
Calculate its moment of inertia.

$$I = \frac{1}{12}ml^2$$
$$I = \frac{1}{12} \times 0.12 \times 0.8^2$$
$$I = 0.0064 \text{ kg m}^2$$

....

3.

A flat disk of pizza dough of mass 800 g and radius 25 cm is spun in a circle about its middle.
Calculate its moment of inertia.

$$I = \frac{1}{2}mr^2$$
$$I = \frac{1}{2} \times 0.8 \times 0.25^2$$
$$I = 0.025 \text{ kg m}^2$$

....

4.

A hula hoop has an axis of rotation as shown. Its mass is 2.0 kg and radius is 60 cm.
Find its moment of inertia.

$$I = mr^2$$
$$I = 2 \times 0.6^2$$
$$I = 0.72 \text{ kg m}^2$$

....

5.

A football of mass 249 g is made to rotate about its centre. The diameter of the ball is 15.0 cm.
Find the ball's moment of inertia about this axis of rotation.

$$I = \frac{2}{5}mr^2$$
$$I = \frac{2}{5} \times 0.249 \times 0.075^2$$
$$I = 5.60 \times 10^{-4} \, \text{kg} \, \text{m}^2$$

....

3.4.1 Combinations of rotating bodies

Let us now consider systems involving more than one rotating body.

Example

A Catherine wheel consists of a light rod 0.64 m in length, pivoted about its centre. At either end of the rod a 0.50 kg firework is attached, each of which provides a force of 25 N perpendicular to the rod when lit. Calculate the angular acceleration of the Catherine wheel when the fireworks are lit.

The angular acceleration is calculated using the equation $T = I\alpha$, so we need to calculate the moment of inertia of the Catherine wheel and the total torque provided by the two fireworks.

Figure 3.8: Catherine wheel diagram

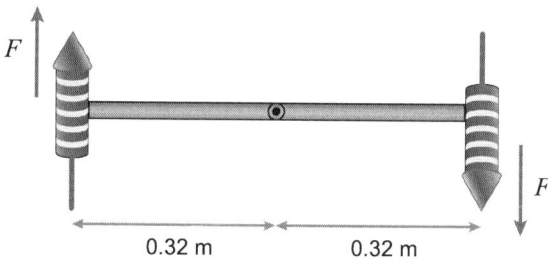

0.32 m   0.32 m

....

The total moment of inertia is

$$I = \sum mr^2$$
$$\therefore I = \left(0.50 \times 0.32^2\right) + \left(0.50 \times 0.32^2\right)$$
$$\therefore I = 0.1024 \, \text{kg} \, \text{m}^2$$

The total torque is the sum of the torques provided by each firework

$$T = \sum Fr$$
$$\therefore T = (25 \times 0.32) + (25 \times 0.32)$$
$$\therefore T = 16 \, \text{N} \, \text{m}$$

The angular acceleration can now be calculated

$$\alpha = \frac{T}{I}$$
$$\therefore \alpha = \frac{16}{0.1024}$$
$$\therefore \alpha = 160 \, \text{rad s}^{-2}$$

. . . .

Combinations of rotating bodies

Different masses are placed on a rotating platform to study the moment of inertia of a system. Investigate how the same moment of inertia can be obtained with different combinations of mass and position.

We will be considering a platform of negligible mass, with an axis through its centre. The radius of the platform is 50 cm.

Q10: What is the moment of inertia when a 1.0 kg mass is placed 40 cm from the axis as shown?

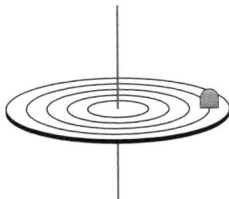

. . . .

Q11: If we remove the 1.0 kg mass, where should a 4.0 kg mass be placed on the platform to produce the same moment of inertia?

. . . .

Q12: Suppose both masses are placed on the platform, with the 4.0 kg mass 30 cm from the centre and the 1.0 kg mass 20 cm from the centre. What is the moment of inertia now?

. . . .

Q13: Calculate the moment of inertia of the following system shown: a 1.0 kg mass 10 cm from the centre, a 2.5 kg mass 20 cm from the centre and a 2.0 kg mass 40 cm from the centre.

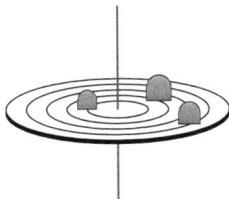

. . . .

Q14: Finally, consider the masses in previous question. If they were combined to form a single object of mass 5.5 kg, where should it be placed to give the same moment of inertia as obtained in question 4?

. . . .

3.5 Summary

Summary

You should now be able to:

- state that the moment (torque) of a force is the tendency of a force to cause or change rotational motion of an object;

- state and apply the equation $T = Fr$;

- state that an unbalanced torque acting on an object produces an angular acceleration;

- state that the moment of inertia of an object is a measure of its resistance to angular acceleration about a given axis;

- explain that the moment of inertia of an object about an axis depends on the mass of the object, and the distribution of the mass about the axis;

- state that the moment of inertia I of a point mass of mass m at a distance r from a fixed axis is given by the equation $I = mr^2$;

- use the relevant equation to calculate the moment of inertia for the following shapes- hoop about centre, rod about centre, rod about end, disc about centre, sphere about centre;

- calculate the moment of inertia of a combination of rotating bodies;

- state and apply the equation $T = I\alpha$.

3.6 Extended information

Web links

There are web links available online exploring the subject further.

. . . .

3.7 Assessment

End of topic 3 test

Go online

The following test contains questions covering the work from this topic.

The following data should be used when required:

Gravitational acceleration on Earth g	$9.8 \, m \, s^{-2}$
Moment of inertia of a point mass	$I = mr^2$
Moment of inertia of a rod about its centre	$I = \frac{1}{12}ml^2$
Moment of inertia of a rod about its end	$I = \frac{1}{3}ml^2$
Moment of inertia of a disc about its centre	$I = \frac{1}{2}mr^2$
Moment of inertia of a sphere about its centre	$I = \frac{2}{5}mr^2$

Q15:

A spanner, length 35 cm, is being used to try to undo a nut stuck on a bolt. A force of 71 N is applied at the end of the spanner.

Calculate the moment which is applied.

Give your answer to *at least* 1 decimal place.

Moment = _____ m

. . . .

Q16:

A torque of 65 Nm is applied to a solid flywheel of moment of inertia 17 kg m^2.

Calculate the resulting angular acceleration.

Give your answer to at least 1 decimal place.

Angular acceleration = _____ rad s^{-2}

. . . .

Q17:

Three objects are arranged on a circular turntable of negligible mass, which is able to rotate about an axis through its centre.

Mass A (0.46 kg) is 20 cm from the axis.

Mass B (0.61 kg) is 40 cm from the axis.

Mass C (0.28 kg) is 70 cm from the axis.

Calculate the total moment of inertia of the loaded turntable about the central axis.

I = _____ kg m^2

. . . .

Q18:

An object of mass 2.9 kg is moving in a circle of radius 1.2 m. Calculate the moment of inertia of the mass about the centre of the circle.

Give your answer rounded to *at least* 1 decimal places.

I = _____ kg m^2

. . . .

Q19:

Masses A (1.5 kg) and B (2 kg) are placed on a platform of negligible mass, which can rotate about an axis XY.

Mass A is positioned 10 cm from XY.

Mass B sits 23 cm from XY.

Calculate the total moment of inertia about the axis XY.

Give your answer rounded to *at least* 2 decimal places.

I = _____ kg m^2

. . . .

Q20:

The moment of inertia of an object is a measure of its resistance to _____ acceleration about a given axis.

The moment of inertia of an object about an axis depends on the _____ of the object, and the distribution of the _____ about the axis.

. . . .

Q21:

A solid cylinder of mass 0.3 kg rolls down a ramp. The diameter of the cylinder is 4.0 cm.

Find the moment of inertia of the cylinder.

I = _____ kg m^2

. . . .

Q22:

The moment of inertia of a CD about a perpendicular axis through its centre is 2.70×10^{-5} kg m^2. The CD has a diameter of 12 cm.

Find its mass in grams.

Give your answer rounded to 3 significant figures.

m = _____ g

. . . .

Q23:

A ball bearing of mass 29.0 g rolls down a slope. The moment of inertia of the ball bearing is 1.05×10^{-6} kg m^2.

Find its radius in cm.

Give your answer rounded to 3 significant figures.

r = _____ cm

. . . .

Topic 4

Angular momentum

Contents

Prerequisite knowledge

- *Understand the relationship between torque, moment of inertia and angular acceleration (Unit 1 - Topic 3).*
- *Understand the principle of conservation of linear momentum.*

Learning objectives

By the end of this topic you should be able to:

- *state that the angular momentum L of a rigid body is given by the equation $L = I\omega$;*
- *use the equation $L = mvr = mr^2\omega$ for a point mass;*
- *state that in the absence of external torques, angular momentum is conserved;*
- *state the expression $E_K = \frac{1}{2}I\omega^2$ for the rotational kinetic energy of a rigid body, and carry out calculations using this relationship.*

4.1   Introduction

In the last topic, we studied the moment of inertia of shapes about different axes of rotation. We will now find the equations for angular momentum and rotational kinetic energy. Again, these compare directly to the linear equations with which you should be familiar.

4.2   Angular momentum

The **angular momentum** L of a body rotating with moment of inertia I and angular velocity ω is defined as

$$L = I\omega$$

(4.1)

....

L is measured in units of kg m^2 s^{-1}. Equation 4.1 is analogous to the expression $p = mv$ for linear momentum, and the same conservation of momentum principle applies. In this case, we can state that the angular momentum of a rotating rigid body is conserved unless an external torque acts on the body.

If we want to describe the angular momentum at a point, then we can substitute for $I = mr^2$ and $\omega = {}^v/_r$ in Equation 4.1

$$L = I\omega \qquad\qquad\qquad L = mr^2 \times {}^v/_r$$
$$\therefore L = mr^2\omega \qquad\qquad \therefore L = mvr$$

4.3   Conservation of angular momentum

The conservation of angular momentum has some interesting consequences. Since angular momentum depends on the moment of inertia, then it depends on the distribution of mass about the axis. If the moment of inertia increases, the angular velocity must decrease to keep the angular momentum constant, and vice versa. This effect can be seen if you spin on a swivel chair. Flinging out your arms and legs moves more of your body mass away from the axis of rotation. Your moment of inertia increases, so your angular velocity must decrease. (This works even better if you have a heavy book in each hand!) If you then hunch up on the chair, your moment of inertia decreases, and hence your angular velocity increases.

4.3.1 Angular momentum in sports

Gymnasts, snowboarders, divers and acrobats all use the same principle. As they move from a stretched to a tucked position, their moment of inertia decreases and this makes their angular velocity increase.

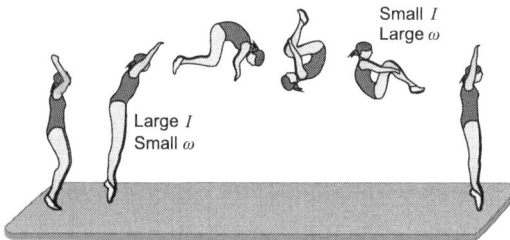

Example

An ice skater is spinning with his arms extended. His angular velocity is 7.0 rad s^{-1}. What is his angular velocity if he pulls his arms in to his sides? The moment of inertia of the skater is 4.7 kg m^2 when his arms are extended, and 1.8 kg m^2 when they are by his sides.

The angular momentum of the skater must be conserved, so the initial angular momentum must equal the final angular momentum.

$$L_i = L_f$$
$$\therefore I_i \omega_i = I_f \omega_f$$
$$\therefore 4.7 \times 7.0 = 1.8 \times \omega_f$$
$$\therefore \omega_f = \frac{4.7 \times 7.0}{1.8}$$
$$\therefore \omega_f = 18 \text{ rad s}^{-1}$$

....

An ice skater

Go online

There is an online activity showing animations of angular momentum of an ice skater.

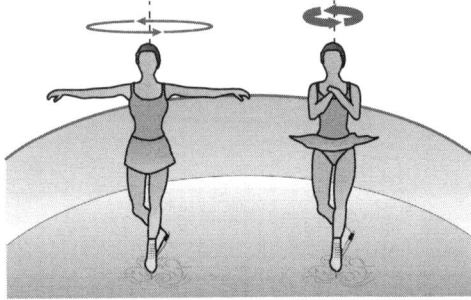

. . . .

4.3.2   Satellite spin

A special device is used to reduce the spin of satellites. It involves two cables wrapped around the satellite, with masses attached to the other end. When activated, the masses are released and move away from the satellite like yoyos, extending to the end of their cables. The moment of inertia therefore increases. In the absence of an external torque, angular momentum must be conserved. So the cord length can be chosen to reduce the angular velocity of the satellite to the desired value. At this point, the masses are then detached.

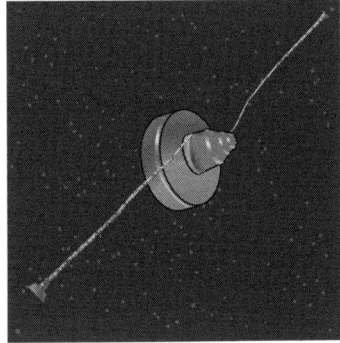

The Mars exploration rover's angular velocity was reduced in this way.

Quiz: Conservation of angular momentum

Q1: A solid object with moment of inertia 2.40 kg m^2 is rotating at 3.00 rad s^{-1}. What is its angular momentum?

Go online

a) 0.800 kg m^2 s^{-1}
b) 1.25 kg m^2 s^{-1}
c) 1.92 kg m^2 s^{-1}
d) 7.20 kg m^2 s^{-1}
e) 17.3 kg m^2 s^{-1}

. . . .

Q2: A disc of mass m and radius r spinning about an axis through its centre has moment of inertia $I = \frac{1}{2}mr^2$. What is the angular momentum of a disc (m = 0.400 kg, r = 0.25 m) spinning at 20 rad s^{-1}?

a) 0.05 kg m^2 s^{-1}
b) 0.25 kg m^2 s^{-1}
c) 5.0 kg m^2 s^{-1}
d) 8.0 kg m^2 s^{-1}
e) 250 kg m^2 s^{-1}

. . . .

Q3: A horizontal turntable is rotating on a frictionless mount at constant angular velocity. What happens if a lump of clay is dropped onto the turntable, sticking to it?

a) The turntable would slow down.
b) The turntable would speed up.
c) The turntable would continue rotating at the same speed.
d) The lump of clay would rotate in the opposite direction to the turntable.
e) The turntable would stop rotating but the lump of clay would continue to rotate.

. . . .

4.4 Rotational kinetic energy

There are also expressions for calculating work done and kinetic energy in rotational motion, that are again analogous to expressions used to describe linear motion. You should be familiar with the expression *work done = force × displacement*. In the case of rotational work, the equivalent expression is

$$\text{work done} = T \times \theta$$

(4.2)

....

In Equation 4.2, T is the torque and θ is the angular displacement. This equation tells us the work done when a torque T applied to an object produces an angular displacement θ of the object.

A spinning body has kinetic energy. The **rotational kinetic energy** of a body is given by

$$\text{Rotational } E_K = \frac{1}{2}I\omega^2$$

(4.3)

....

Note that this energy is due to the rotation about an axis. If we consider a wheel spinning about its axis, then this is the total kinetic energy. What if the wheel is rolling along the ground? In this case the wheel has both rotational kinetic energy, due to its turning motion, and translational kinetic energy due to its moving along the ground. The total kinetic energy in this case is

$$\text{Total } E_K = \text{Rotational } E_K + \text{ Translational } E_K$$
$$\therefore \text{ Total } E_K = \frac{1}{2}I\omega^2 + \frac{1}{2}mv^2$$

Note that the translational speed is the same as the tangential speed at the circumference of the wheel, so long as the wheel does not slip. If you cannot see why, imagine the wheel rotating through one complete circle. A point on the circumference moves through a distance $2\pi r$ in one revolution. The wheel must have rolled the same distance, and it has done so in the same period of time. The translational speed must therefore be equal to the tangential speed at the circumference.

Example

A solid sphere of radius r and mass m has moment of inertia $I = \frac{2}{5}mr^2$ about any axis though its centre. Calculate the translational and rotational kinetic energies of a sphere of mass 0.40 kg and radius 20 cm rolling (without slipping) along a horizontal surface with translational speed 5.0 m s^{-1}.

The translational kinetic energy is

$$\text{Translational } E_K = \frac{1}{2}mv^2$$
$$\therefore \text{Translational } E_K = \frac{1}{2} \times 0.40 \times 5.0^2$$
$$\therefore \text{Translational } E_K = 5.0 \text{ J}$$

To find the rotational kinetic energy, we need to find the moment of inertia I and angular velocity ω of the sphere. If the sphere has a translational speed of v m s^{-1}, then a point on the circumference of the sphere must also have speed v. The angular velocity is therefore equal to $v/_r$, which in this case is

$$\omega = \frac{v}{r} = \frac{5.0}{0.20} = 25 \text{ rad s}^{-1}$$

The rotational kinetic energy is

$$\text{Rotational } E_K = \frac{1}{2}I\omega^2$$
$$\therefore \text{Rotational } E_K = \frac{1}{2} \times \frac{2}{5}mr^2 \times \omega^2$$
$$\therefore \text{Rotational } E_K = \frac{1}{5} \times 0.40 \times 0.20^2 \times 25^2$$
$$\therefore \text{Rotational } E_K = 2.0 \text{ J}$$

. . . .

Flywheels

Flywheels are spinning wheels or discs with a fixed axle so that rotation is only about one axis. They can be used to store energy in machines, such as push and go toy cars. Some experimental vehicles use flywheels to charge up their batteries. Your teacher may ask you to carry out a practical experiment to find the moment of inertia of a flywheel.

Total energy of a rolling body

At this stage there is an online activity allowing you to plot out the energy changes that occur for different objects rolling down a slope..

Go online

. . . .

Quiz: Angular momentum and rotational kinetic energy

Go online

Q4: A square sheet of mass m and side length I has a moment of inertia $I = \frac{1}{3}ml^2$ when rotating about one edge. What is the rotational kinetic energy of a square sheet of mass 1.0 kg and side 0.50 m rotating at 0.60 rad s^{-1} about one edge?

a) 9.0×10^{-4} J
b) 4.2×10^{-3} J
c) 0.015 J
d) 0.030 J
e) 0.18 J

....

Q5: A rotating platform is spinning at a rate of 1.50 revolutions per second. If the moment of inertia of the platform is 4.20 kg m^2, what is the rotational kinetic energy of the platform?

a) 4.73 J
b) 19.8 J
c) 39.6 J
d) 47.3 J
e) 187 J

....

4.5   Comparing linear and angular motion

Using the moment of inertia allows us to describe the angular motion of rigid bodies in the same way that we describe linear motion of point objects. The same conservation rules apply, and the following tables should clarify the different quantities and relationships:

Table 4.1: Kinematic relationships

Linear motion	Angular motion
$v = u + at$	$\omega = \omega_0 + \alpha t$
$v^2 = u^2 + 2as$	$\omega^2 = \omega_0{}^2 + 2\alpha\theta$
$s = ut + \frac{1}{2}at^2$	$\theta = \omega_0 t + \frac{1}{2}\alpha t^2$

....

Table 4.2: Linear and angular motion

Linear motion	Angular motion
Displacement s	Angular displacement θ
Force F	Torque T
Velocity v	Angular velocity ω
Acceleration a	Angular acceleration α

. . . .

Table 4.3: Newton's second law and kinetic energy

Linear motion	Angular motion
Newton's second law $$F = ma$$	Newton's second law $$T = I\alpha$$
Momentum $$p = mv$$	Angular momentum $$L = I\omega$$
Work done $$W = Fs$$	Work done $$W = T\theta$$
Translational kinetic energy Translational $E_K = \dfrac{1}{2}mv^2$	Rotational kinetic energy Rotational $E_K = \dfrac{1}{2}I\omega^2$

. . . .

The moment of inertia of a rigid body depends on the mass of the body, and how that mass is distributed about the axis of rotation. The axis must be specified when a moment of inertia is being calculated. For a collection of point masses, the total moment of inertia is the sum of the individual moments.

4.6   Summary

Summary

You should now be able to:

- state that the angular momentum L of a rigid body is given by the equation $L = I\omega$;

- use the equation $L = mvr = mr^2\omega$ for a point mass;

- state that in the absence of external torques, angular momentum is conserved;

- state the expression $E_K = \frac{1}{2}I\omega^2$ for the rotational kinetic energy of a rigid body, and carry out calculations using this relationship.

4.7   Extended information

Web links

There are web links available online exploring the subject further.

. . . .

4.8   Assessment

End of topic 4 test

The following test contains questions covering the work from this topic.

Go online

The following data should be used when required:

Gravitational acceleration on Earth g	$9.8\ m\ s^{-2}$
Moment of inertia of a point mass	$I = mr^2$
Moment of inertia of a rod about its centre	$I = \frac{1}{12}ml^2$
Moment of inertia of a rod about its end	$I = \frac{1}{3}ml^2$
Moment of inertia of a disc about its centre	$I = \frac{1}{2}mr^2$
Moment of inertia of a sphere about its centre	$I = \frac{2}{5}mr^2$

Q6:

A solid sphere, radius 20.0 cm and mass 1.36 kg, is rotating about an axis through its centre with angular velocity 4.65 rad s^{-1}.

Calculate the rotational kinetic energy of the sphere.

Rotational E_K = _____ J

. . . .

Q7:

A flat horizontal disc of moment of inertia 1.2 kg m^2 is rotating at 4.5 rad s^{-1} about a vertical axis through its centre.

A 0.13 kg mass is dropped onto the disc, landing without slipping 1.7 m from the centre.

Calculate the new angular velocity of the disc.

ω = _____ rad s^{-1}

. . . .

Q8:

A solid cylinder of radius 0.76 m and mass 7.9 kg is at rest. A 4 N m torque is applied to the cylinder about an axis through its centre.

Calculate the angular velocity of the cylinder after the torque has been applied for 2.0 s.

ω = _____ rad s^{-1}

. . . .

Q9:

A solid cylinder of mass 3.5 kg and radius 1.2 m is rolling without slipping along a horizontal road.

The translational speed of the cylinder is 5.4 m s^{-1}.

1. Calculate the angular velocity of the cylinder.

   ω = _____ rad s^{-1}

2. Calculate the total kinetic energy of the cylinder.

   Total E_K = _____ J

. . . .

Topic 5

Gravitation

Contents

Prerequisite knowledge

- *Newton's laws of motion.*
- *Circular motion - centripetal force, periodic time.*

Learning objectives

By the end of this topic you should be able to:

- *state and apply the equation $F\,\frac{GMm}{r^2}$ to calculate the gravitational force between two objects;*
- *calculate the weight of an object using Newton's Universal Law of Gravitation;*
- *state what is meant by a gravitational field, and calculate the gravitational field strength at a point in the field;*

- calculate the value of the acceleration due to gravity at a point in a gravitational field, given the local conditions;

- sketch the field lines around a planet and a planet-moon system;

- use appropriate relationships to carry out calculations involving the period of a satellite in circular orbit, a satellite's mass, speed and orbital radius;

- explain what is meant by a geostationary satellite;

- state the expression $V = \frac{-GM}{r}$ and use it to calculate the gravitational potential V at a point in a gravitational field;

- calculate the gravitational potential energy of a mass in a gravitational field and calculate the change in the potential energy when a mass is moved between points in the field;

- define escape velocity as the minimum velocity required to allow a mass to escape a gravitational field;

- derive the expression $v = \sqrt{\frac{2GM}{r}}$ and use it to calculate the escape velocity.

5.1 Introduction

You studied Newton's Universal Law of Gravitation at Higher. In this topic we will look at how this law can be used to establish a relationship between gravitational field strength and the height above the Earth. Throughout this topic, some approximations will be made with regard to planets and their orbits. It will be assumed that the Sun, the Moon, the Earth and other planets are all spherical objects and that all orbits are circular. We will see how the centripetal force on a satellite due to the gravitational attraction determines the speed and period of the satellite.

The second half of the topic explores gravitational fields. We will look at the potential energy of a body in a gravitational field, so that the work done in moving a body around in a gravitational field can be calculated. Finally, we will consider the **escape velocity** of a rocket, which is how fast it needs to be travelling when it takes off in order to escape a body's gravitational field. This will help us understand black holes in a future topic.

5.2 Newton's law of gravitation

Working in the 17th century, Sir Isaac Newton discovered the **Universal Law of Gravitation**. He used his own observations, along with those of Johannes Kepler, who had formulated a set of laws that described the motion of the planets around the Sun. Kepler's laws will be studied in the next topic.

Newton's Law of Gravitation states that there is a force of attraction between any two objects in the universe. The size of the force is proportional to the product of the masses of the two objects, and inversely proportional to the square of the distance between them. This law can be summed up in the equation

$$F = \frac{Gm_1m_2}{r^2}$$

(5.1)

. . . .

In Equation 5.1, m_1 and m_2 are the masses of the two objects, and r is the distance between them. The constant of proportionality is G, the Universal constant of gravitation. The value of G is 6.67 × 10⁻¹¹ m³ kg⁻¹ s⁻². A simple example will show us the order of magnitude of this force.

Example

Consider two point masses, each 2.00 kg, placed 1.20 m apart on a table top. Calculate the magnitude of the gravitational force between the two masses.

Using Newton's Law of Gravitation

$$F = \frac{Gm_1m_2}{r^2}$$
$$\therefore F = \frac{6.67 \times 10^{-11} \times 2.00 \times 2.00}{1.20^2}$$
$$\therefore F = \frac{2.668 \times 10^{-10}}{1.44}$$
$$\therefore F = 1.85 \times 10^{-10} \text{ N}$$

The gravitational force between these two masses is only 1.85×10^{-10} N. This is an extremely small force, one which is not going to be noticeable in everyday life.

. . . .

We rarely notice the gravitational force that exists between everyday objects as it is such a small force. You do not have to fight against gravity every time you walk past a large building, for example, as the gravitational force that the building exerts on you is too small to notice. Because the constant G in Newton's Law of Gravitation is so small, the gravitational force between everyday objects is usually negligible. The force only really becomes important when we are dealing with extremely large masses such as planets.

The gravitational force is always attractive, and always acts in the direction of the straight line joining the two objects. According to Newton's third law of motion, the gravitational force is exerted on **both** objects. As the Earth exerts a gravitational force on you, so you exert an equal force on the Earth.

Most of the work we will be doing on gravitation concerns the forces acting between planets and stars. So far we have only considered point objects, so do we need to adapt Newton's Law of Gravitation when we are dealing with larger bodies? The answer is no - for spherical objects (or more accurately, objects with a spherically symmetric mass distribution), the gravitational interaction is exactly the same as it would be if all the mass was concentrated at the centre of the sphere. Remember, we will assume in all our calculations that the planets and stars we are dealing with are spherical.

Example

The Earth has a radius of 6.4×10^6 m and a mass 6.0×10^{24} kg. What is the gravitational force due to the Earth acting on a woman of mass 60.0 kg standing on the surface of the Earth?

The solution is obtained by calculating the force between two point objects placed 6.4×10^6 m apart, if we treat the Earth as a uniform sphere. Hence

$$F = \frac{Gm_1m_2}{r^2}$$

$$\therefore F = \frac{6.67 \times 10^{-11} \times 6.0 \times 10^{24} \times 60.0}{(6.4 \times 10^6)^2}$$

$$F = \frac{2.4012 \times 10^{16}}{4.096 \times 10^{13}}$$

$$F = 590 \, \text{N}$$

This answer has been rounded to two significant figures since the mass of the Earth was only stated to two significant figures.

The gravitational force acting on the woman is 590 N, and this force is directed towards the centre of the Earth.

If you calculate the force due to gravity acting on the woman by another method, using $F = m \times g$, you should also get the answer 590 N, when rounded to two significant figures.

. . . .

5.2.1 The Cavendish-Boys experiments to determine G

An accurate measurement of G can be carried out using the Cavendish-Boys method. Cavendish first performed this experiment in the late 18th century and the accurate determination was performed nearly 100 years later by Boys. The experiment uses a **torsion balance** to measure the gravitational force between lead spheres.

Figure 5.1: Cavendish Boys-experiment, viewed from (a) the side, and (b) the top

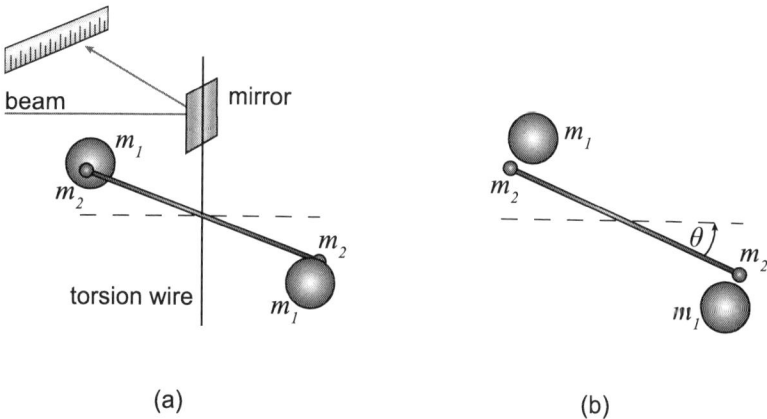

(a)   (b)

. . . .

Two identical masses (m_1) are held close to two smaller masses (m_2). These smaller masses are attached to the ends of a light rod of length l suspended from a torsion wire in a draft-free chamber. By reflecting a narrow beam of light from a mirror attached to the torsion wire, the angular deflection of the wire can be measured as the masses m_1 are brought close to the masses m_2. In equilibrium, the rotation caused by the gravitational force between m_1 and m_2 is balanced by the restoring torque (turning force) in the wire. The equipment must be calibrated first to determine a quantity called the torsional constant c of the wire. The restoring torque is then equal to $c \times \theta$ when the angular displacement of the rod is θ rad (see Figure 5.1(b)). Hence in equilibrium,

Torque due to gravitational force $=$ Restoring torque

$$\therefore G\frac{m_1 m_2}{r^2}l = c\theta$$

(5.2)

. . . .

5.2.2 Maskelyne's experiment

One of the consequences of Newton's Universal Law of Gravitation is that on level ground a pendulum ought to hang vertically, since it is attracted towards the centre of the Earth. However, if a mountain is nearby, then its gravitational attraction will pull the pendulum slightly off vertical. For this reason, in the late 18th century, an astronomer called Maskelyne used a pendulum and a Scottish mountain called Schiehallion to estimate the mass of the Earth. He chose to use this Scottish mountain since it was in an isolated location and its regular shape allowed its mass to be estimated.

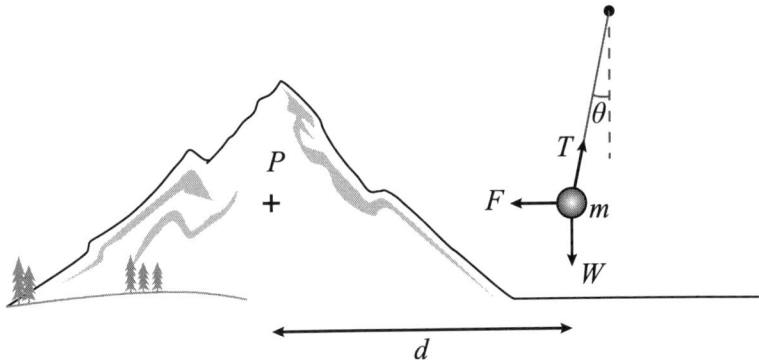

Two gravitational forces act on the pendulum bob of mass m. They are the gravitational attraction of Schiehallion and the bob's weight. They can be found from

$$F = \frac{GM_{\text{Schiehallion}}m}{d^2}$$

and

$$W = \frac{GM_{\text{Earth}}m}{r_E{}^2}$$

where d is the distance between the mountain and the bob. As usual, r_E is the radius of the Earth.

Maskelyne knew that the vertical component of the tension T must balance the bob's weight. He knew the horizontal component of the tension T must balance the force exerted by Schiehallion. Therefore, the angle θ of the pendulum to the vertical could be expressed by the equation

$$\tan\theta = \frac{\frac{GM_{\text{Schiehallion}}m}{d^2}}{\frac{GM_{\text{Earth}}m}{r_E{}^2}}$$

This meant that, by measuring the angle of the pendulum to the vertical, Maskelyne was then able to solve for the mass of the Earth.

5.3 Weight

The **weight** W of an object of mass m can be defined as the gravitational force exerted on it by the Earth.

$$W = F_{\text{grav}} = \frac{GM_E m}{r_E{}^2}$$

(5.3)

. . . .

In Equation 5.3, M_E and r_E are the mass and radius of the Earth.

The acceleration due to gravity, g, is found from Newton's second law of motion

$$F = ma$$
$$\therefore W = mg$$

We can substitute for F in this equation

$$mg = \frac{GM_E m}{r_E{}^2}$$
$$\therefore g = \frac{GM_E}{r_E{}^2}$$

(5.4)

. . . .

So the acceleration of an object due to gravity close to the Earth's surface does not depend on the mass of the object. In the absence of friction, all objects fall with the same acceleration.

Equation 5.4 is a specific equation for calculating g on the Earth's surface. In general, at a distance r from the centre of a body (a star or a planet, say) of mass M, the value of g is given by

$$g = \frac{GM}{r^2}$$

(5.5)

. . . .

The mass of an object is constant; it is an intrinsic property of that object. The weight of an object tells us the magnitude of the gravitational force acting upon it, so it is not a constant.

Example

Compare the values of g on the surfaces of the Earth (M_E = 6.0 × 10²⁴ kg, r_E = 6.4 × 10⁶ m) and the Moon (M_M = 7.3 × 10²² kg, r_M = 1.7 × 10⁶ m).

To solve this problem, use Equation 5.5 with the appropriate values

$$g = \frac{GM_E}{r_E{}^2}$$
$$\therefore g = \frac{6.67 \times 10^{-11} \times 6.0 \times 10^{24}}{\left(6.4 \times 10^6\right)^2}$$
$$\therefore g = 9.8 \text{ m s}^{-2}$$

$$g = \frac{GM_M}{r_M{}^2}$$
$$\therefore g = \frac{6.67 \times 10^{-11} \times 7.3 \times 10^{22}}{\left(1.7 \times 10^6\right)^2}$$
$$\therefore g = 1.7 \text{ m s}^{-2}$$

This gives a value for g on the surface of the Earth of 9.8 m s^{-2}, compared to a value on the surface of the Moon of 1.7 m s^{-2}. The value for g on the surface of the Moon is usually quoted as 1.6 m s^{-2}, though it varies over its entire surface by about 0.03 ms^{-2}. The value calculated here differs since the mass and radius of the Moon were only quoted to two significant figures in the question.

. . . .

Quiz: Gravitational force

Useful data:

Go online

Universal constant of gravitation G	$6.67 \times 10^{-11}\ N\ m^2\ kg^{-2}$
Mass of the Moon M_M	$7.3 \times 10^{22}\ kg$
Radius of the Moon r_M	$1.7 \times 10^6\ m$
Mass of Venus M_V	$4.87 \times 10^{24}\ kg$
Radius of Venus r_V	$6.05 \times 10^6\ m$

Q1: Two snooker balls, each of mass 0.25 kg, are at rest on a snooker table with their centres 0.20 m apart. What is the magnitude of the gravitational force that exists between them?

a) 2.1×10^{-11} N
b) 4.3×10^{-11} N
c) 8.3×10^{-11} N
d) 1.0×10^{-10} N
e) 4.2×10^{-10} N

. . . .

Q2: The Sun exerts a gravitational force F_S on the Earth. The Earth exerts a gravitational force F_E on the Sun. Which one of these statements about the magnitudes of F_S and F_E is true?

a) $F_S = F_E$
b) $F_S < F_E$
c) $F_S > F_E$
d) $F_S / F_E = \text{mass}_S / \text{mass}_E$
e) $F_S / F_E = (\text{mass}_S)^2 / (\text{mass}_E)^0$

. . . .

Q3: What is the weight of a 5.00 kg mass placed on the surface of the Moon?

a) 0.12 N
b) 1.4 N
c) 1.6 N
d) 8.4 N
e) 49 N

....

Q4: An object is taken from sea level to the top of Mount Everest. Which one of the following statements is true?

a) Its mass remains constant but its weight increases.
b) Its mass remains constant but its weight decreases.
c) Its weight remains constant but its mass increases.
d) Its weight remains constant but its mass decreases.
e) Neither its mass nor its weight alter.

....

Q5: What is the value of the acceleration due to gravity on the surface of the planet Venus?

a) 0.887 m s^{-2}
b) 5.37 m s^{-2}
c) 6.67 m s^{-2}
d) 8.29 m s^{-2}
e) 8.87 m s^{-2}

....

5.4 Gravitational fields

The region of space around an object A, in which A exerts a gravitational force on another object B placed in that region, is called the **gravitational field** of A. The concept of a field is used in many situations in physics, such as the electric field surrounding a charged particle, or the magnetic field around a bar magnet.

The **gravitational field strength** g at a point in a gravitational field is defined as the gravitational force acting on a unit mass placed at that point in the field. At a distance r from a point object of mass M, the gravitational field strength g is given by

$$g = \frac{GM}{r^2}$$

(5.6)

....

The units of g are N kg^{-1}. These units are equivalent to m s^{-2}, so the gravitational field strength at a point in a gravitational field is equal to the acceleration due to gravity at that point, and Equation 5.6 is identical to Equation 5.5. This is why the same symbol g is used for both the gravitational field strength and the acceleration due to gravity.

Example

The Earth orbits the Sun with a mean radius of 1.5×10^{11} m. What is the gravitational field strength on Earth due to the Sun, given that the mass of the Sun is 2.0×10^{30} kg.

In the Sun's gravitational field, the field strength is given by

$$g = \frac{GM_S}{r^2}$$

So, at the location of the Earth, the field strength is

$$g = \frac{GM_S}{r^2}$$
$$\therefore g = \frac{6.67 \times 10^{-11} \times 2.0 \times 10^{30}}{\left(1.5 \times 10^{11}\right)^2}$$
$$\therefore g = 5.9 \times 10^{-3} \text{ N kg}^{-1}$$

....

The gravitational field around a body is often shown diagramatically by drawing field lines. Figure 5.2 shows the gravitational field lines around a point object. The pattern of the field lines is symmetrical in three dimensions. This pattern would be exactly the same outside an object with a spherically symmetric mass distribution. The lines are symmetrical about the centre of the object and show the direction of the force exerted on any mass placed in the field. The closer together the field lines are, the greater the field strength.

Figure 5.2: Gravitational field lines around a point mass

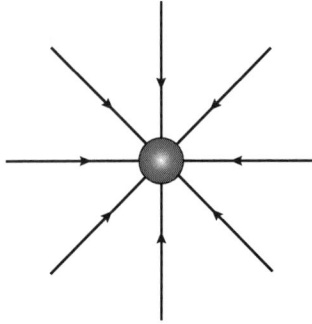

. . . .

The gravitational field is an example of a **conservative field**. This is a field in which the work done in moving from one point to another in the field is independent of the path taken.

The field lines representing the gravitational field caused by two or more objects can also be sketched. The gravitational field due to two point masses is shown in Figure 5.3.

Figure 5.3: Gravitational field lines around two identical point masses

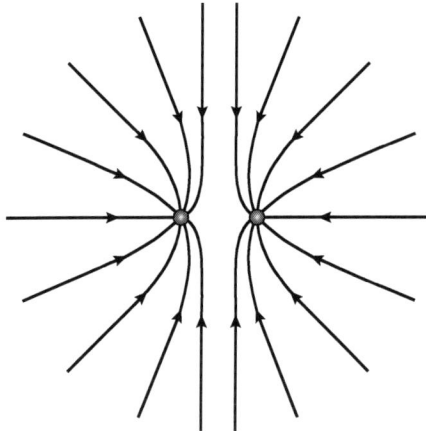

. . . .

To calculate the gravitational field strength at any point in the field due to two point masses or spheres, we take the vector sum of the two individual fields.

Example

The mean distance between the centre of the Earth and the centre of the Moon is 3.84×10^8 m. Given that the mass of the Earth is 6.0×10^{24} kg and the mass of the Moon is 7.3×10^{22} kg, at what distance from the centre of the Earth is the point where the total gravitational field strength due to the Earth and the Moon is zero?

A sketch is useful in solving this sort of problem, as it can clearly show the direction of the field vectors.

Figure 5.4: Earth - Moon system

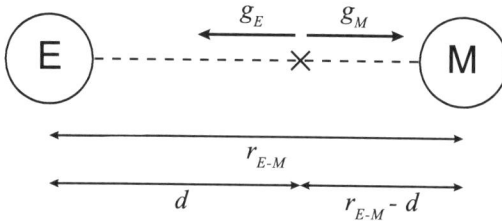

At position X, the point where the total gravitational field strength is zero, the gravitational field strength g_E of the Earth is balanced by g_M, the gravitational field strength of the Moon. Position X is a distance d from the centre of the Earth. To find the distance d, we must solve the equation $g_E = g_M$

$$g_E = g_M$$
$$\therefore \frac{GM_E}{d^2} = \frac{GM_M}{(r_{E-M} - d)^2}$$
$$\therefore \frac{M_E}{d^2} = \frac{M_M}{(r_{E-M} - d)^2}$$
$$\therefore \frac{\sqrt{M_E}}{d} = \frac{\sqrt{M_M}}{(r_{E-M} - d)}$$
$$\therefore \sqrt{M_E} \times (r_{E-M} - d) = \sqrt{M_M} \times d$$
$$\therefore \sqrt{M_E} \times r_{E-M} = d \times \left(\sqrt{M_M} + \sqrt{M_E}\right)$$
$$\therefore 2.449 \times 10^{12} \times 3.84 \times 10^8 = d \times \left(2.702 \times 10^{11} + 2.449 \times 10^{12}\right)$$
$$\therefore d = \frac{9.404 \times 10^{20}}{2.719 \times 10^{12}}$$
$$\therefore d = 3.5 \times 10^8 \text{ m}$$

The combined gravitational field strength is zero at a point 3.5×10^8 m from the centre of the Earth. This point is very much closer to the Moon than the Earth because the Earth is much more massive than the Moon.

Gravitational fields

Go online

At this stage there is an online activity where you can see the pattern of the gravitational field lines around a single object or a collection of objects.

The gravitational field lines around an isolated object are plotted. A second object can be added to the system, and the field lines are re-plotted if the separation between the objects or their masses are changed.

With this simulation you can see the pattern of the gravitational field lines around a single object or a collection of objects. Note that masses are always added at the same position, so if you do not move the masses they will all be located at the same point.

. . . .

The Earth - Moon system

The last interactivity allowed us to explore the gravitational field line pattern around two masses. The gravitational field line pattern for a planet-moon system is therefore as follows.

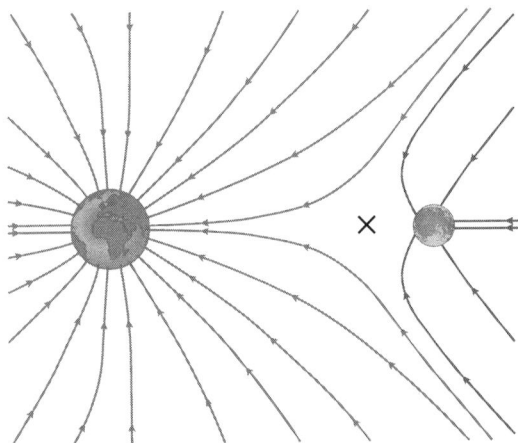

Note the gravitational field lines meet the surface of the Earth and the Moon at 90 ° to their surface. The point at which the gravitational field strength is zero has been labelled X. We saw earlier that this is closer to the Moon than the Earth because the Earth has a much larger mass.

Quiz: Gravitational fields

Go online

Useful data:

Universal constant of gravitation G	6.67×10^{-11} m^3 kg^{-1} s^{-2}

Q6: The planet Jupiter has a mass of 1.90×10^{27} kg and a radius of 6.91×10^7 m.

What is the gravitational field strength on the surface of Jupiter?

a) 0.040 N kg^{-1}
b) 1.83 N kg^{-1}
c) 9.81 N kg^{-1}
d) 26.5 N kg^{-1}
e) 168 N kg^{-1}

. . . .

Q7: Which one of the following units is equivalent to the units used to express gravitational field strength?

a) m s^{-2}
b) N m s^{-2}
c) kg m^{-2}
d) kg N^{-2}
e) N s^{-2}

. . . .

Q8: The gravitational field strength at a distance 5.0×10^5 m from the centre of a planet is 8.0 N kg^{-1}. At what distance from the centre of the planet is the field strength equal to 4.0 N kg^{-1}?

a) 7.1×10^5 m
b) 9.0×10^5 m
c) 1.0×10^6 m
d) 2.0×10^6 m
e) 2.5×10^{11} m

. . . .

Q9: P is a point mass (mass $9m$) and Q is a point mass (mass m). P and Q are separated by 4.0 m. At what point on the line joining P to Q is the net gravitational field strength zero?

a) 0.44 m from P
b) 1.33 m from P
c) 3.0 m from P
d) 3.6 m from P
e) 3.95 m from P

. . . .

Q10: Astronomical observations tell us that the radius of the planet Neptune is 2.48×10^7 m, and the gravitational field strength at its surface is 11.2 N kg^{-1}. What is the mass of Neptune?

a) 3.66×10^3 kg
b) 4.59×10^4 kg
c) 4.16×10^{18} kg
d) 1.03×10^{26} kg
e) 1.16×10^{27} kg

. . . .

5.5   Satellite motion

Man-made satellites orbiting the Earth are used extensively by the telecommunications industry. Other artificial satellites are used for applications such as meteorological observations. The Earth does have a natural satellite too, of course, namely the Moon. In this section we will be studying the motion of satellites, investigating the relationship between orbit radius and orbit period.

5.5.1   Satellite motion

The following interactive activity provides a good starting point for an investigation of satellite motion. A projectile fired from a high mountain will fall to Earth. If the mountain is high enough or the projectile has a great enough speed, the curvature of the Earth must be taken into account when trying to predict where the projectile will land.

Newton's cannon

Go online

At this stage there is an online activity where you can see the initial launch velocity that will enable a constant orbit around the Earth.

. . . .

If we ignore air resistance, then the only force acting on a projectile is the gravitational force exerted on it by the Earth. If the speed of the projectile is increased, the projectile travels further and further horizontally before hitting the ground, until eventually it does not hit the ground at all, but arrives back at its starting point having travelled in a circular path around the Earth. This is the motion of a satellite in a circular orbit around the Earth. We can think of the satellite as 'falling' under the influence of gravity. If the speed is increased still further, with the projectile being released from the same point, it will move in an elliptical orbit around the Earth. Even greater initial speed, and the projectile will reach escape velocity, which is discussed later in this Topic.

If a satellite is orbiting the Earth with a uniform speed, then there must be a constant centripetal force acting on it. We have seen that the only force acting on the satellite (mass M) is the gravitational force exerted on it by the Earth (mass M_E). This force acts towards the centre of the Earth and is the centripetal force. Thus

$$\frac{GM_E m}{r^2} = \frac{mv^2}{r}$$
$$\therefore v^2 = \frac{GM_E}{r}$$
$$\therefore v = \sqrt{\frac{GM_E}{r}}$$

(5.7)

. . . .

Equation 5.7 shows that the speed of the satellite is determined by the radius of the satellite's orbit. The inverse relationship means that the speed increases when the orbit radius decreases. The mass of the satellite does not affect the speed. It is this fact that leads to the effect of 'weightlessness' experienced by astronauts. An astronaut in a satellite can be considered to be a satellite himself, and will be orbiting at the same speed as the satellite. Since there will be no force pushing the astronaut against the floor or the walls, the astronaut experiences an apparent weightlessness relative to the satellite. A common misconception is that the weightlessness an astronaut feels is due to a lack of gravity. In fact, it is because the astronaut and the satellite are both 'falling' with the same acceleration.

The period T of an object performing circular motion with radius r is related to the speed v of the object by the equation

$$v = \frac{2\pi r}{T}$$

(You should remember that $2\pi r$ is the circumference of a circle.) Substituting this expression into Equation 5.7, we can derive an expression for the period T of a satellite orbiting the Earth.

$$v = \sqrt{\frac{GM_E}{r}}$$
$$\therefore \frac{2\pi r}{T} = \sqrt{\frac{GM_E}{r}}$$
$$\therefore \frac{T}{2\pi r} = \sqrt{\frac{r}{GM_E}}$$
$$\therefore T = 2\pi \sqrt{\frac{r^3}{GM_E}}$$

(5.8)

. . . .

Again, the periodic time does not depend on the mass of the satellite.

Example

A meteorological satellite orbits the Earth at an altitude of 2.50×10^5 m above the Earth's surface. What are the speed and periodic time of the satellite?

Before we can calculate the speed, we have to find the radius of the satellite's orbit, which is equal to the radius of the Earth plus the altitude of the satellite above the Earth's surface.

$$r = r_E + \text{altitude}$$
$$\therefore r = \left(6.4 \times 10^6\right) + \left(2.50 \times 10^5\right)$$
$$\therefore r = 6.65 \times 10^6 \text{ m}$$

The speed can now be calculated using Equation 5.7.

$$v = \sqrt{\frac{GM_E}{r}}$$
$$\therefore v = \sqrt{\frac{6.67 \times 10^{-11} \times 6.0 \times 10^{24}}{6.65 \times 10^6}}$$
$$\therefore v = 7.8 \times 10^3 \text{ m s}^{-1}$$

To calculate the period T, we can put this value of v into the circular motion equation.

$$v = \frac{2\pi r}{T}$$
$$\therefore T = \frac{2\pi r}{v}$$
$$\therefore T = \frac{2\pi \times 6.65 \times 10^6}{7.8 \times 10^3}$$
$$\therefore T = 5.4 \times 10^3 \text{ s}$$

The satellite orbits with a speed of 7.8 $\times$ 10^3 m s^{-1} and a period of 5400 s (90 minutes).

. . . .

The total energy of a satellite in orbit around a planet is made up of kinetic energy due to its motion and potential energy due to its position in the gravitational field of the planet. The following activity will show how these two components of the total energy change if the orbit of the satellite changes.

Changing the orbit of a satellite

There is an online simulation which shows how the energy of a satellite changes when its orbit changes.

Go online

. . . .

5.5.2 Geostationary satellites

A satellite orbiting the Earth above the equator with a periodic time exactly equal to one day will appear to be fixed above that point on the Earth. That is to say, the satellite will always appear to be in the same position in the sky to an observer on Earth. Such a satellite is called a **geostationary satellite**. Communications satellites are all in geostationary orbits, so that a fixed receiver dish back on Earth can receive a signal. A geostationary satellite must be orbiting the Earth with the same angular velocity as the Earth's rotation about its axis. This orbit is sometimes referred to as a **parking orbit**.

The angular velocity ω is equal to $2\pi/T$. A satellite of mass m in a geostationary orbit of radius r around the Earth requires a centripetal force given by the equation

$$F = mr\omega^2$$
$$\therefore F = mr \times \left(\frac{2\pi}{T}\right)^2$$
$$\therefore F = \frac{4\pi^2 mr}{T^2}$$

This centripetal force is provided by the gravitational force exerted by the Earth on the satellite, hence the centripetal force is equal to the gravitational force calculated using Newton's Law of Gravitation.

$$\frac{4\pi^2 mr}{T^2} = \frac{GM_E m}{r^2}$$
$$\therefore r^3 = \frac{GM_E T^2}{4\pi^2}$$
$$\therefore r = \sqrt[3]{\frac{GM_E T^2}{4\pi^2}}$$

(5.9)

. . . .

(Note that Equation 5.9 is equivalent to Equation 5.8 rearranged to make r the subject of the equation).

We can evaluate the orbital radius of a geostationary satellite as follows. The Earth rotates about its axis every 24 hours, so the periodic time T of this motion is (24 × 60 × 60) = 86400 s. Since G and M_E are constants, the radius of the geostationary orbit is fixed. Putting the values of T, G and M_E into Equation 5.9, the radius of orbit equals 4.23 × 10^7 m. The radius of the Earth is 6.4 × 10^6 m, so a geostationary satellite orbits (4.23 × 10^7) - (6.4 × 10^6) = 3.6 × 10^7 m above the Earth's surface.

5.5.3 Kepler's laws

The three laws that govern the motion of planets were discovered by Johannes Kepler in the early 17th century. The laws were based on the astronomical observations of Tycho Brahe, and are known as Kepler's laws. These laws state that:

1. The planets move in elliptical orbits, with the Sun at one focus of the ellipse.

2. The line joining the Sun and a planet sweeps out equal areas in equal times.

3. For each planet, the square of the periodic time of its orbit is proportional to the cube of its mean distance from the Sun.

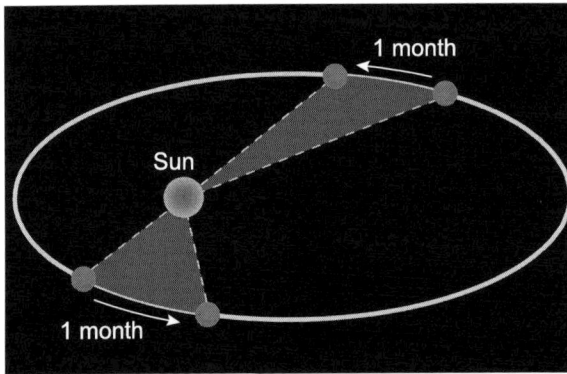

In the simple analysis we have carried out in this Topic, we have assumed circular orbits, which is a special case of the first two laws. The mathematics of ellipses and elliptical motion are beyond the scope of this course. For uniform circular motion, the second law should be obvious to you. We have already encountered the third law in Equation 5.8. Newton also made use of the third law when deriving his universal law of gravitation; he suspected that the gravitational force between two bodies had an inverse-square dependence on the distance between the bodies. Making this assumption, he then showed that this was consistent with Kepler's third law.

Quiz: Satellite motion

Go online

Useful data:

Universal constant of gravitation G	$6.67 \times 10^{-11} \ m^3 \ kg^{-1} \ s^{-2}$
Mass of the Earth M_E	$6.0 \times 10^{24} \ kg$
Radius of the Earth r_E	$6.4 \times 10^6 \ m$

Q11: A satellite of mass 1200 kg is orbiting the Earth with an orbit radius of 6.80×10^6 m. What is the speed of the satellite?

a) 490 m s^{-1}
b) 7700 m s^{-1}
c) 1.03×10^4 m s^{-1}
d) 5.9×10^7 m s^{-1}
e) 2.9×10^9 m s^{-1}

....

Q12: Which one of the following statements is *false*?

a) The speed of a satellite orbiting a planet depends on the mass of the planet.
b) The period of a satellite orbiting a planet depends on the mass of the planet.
c) Two satellites of different mass, orbiting the same planet with the same orbit radius, will have the same period.
d) A satellite's speed has an inverse-square-root dependence on its radius of orbit.
e) The period of a satellite orbiting the Earth depends on the mass of the satellite.

....

Q13: A geostationary satellite orbits at a height of 3.58×10^7 m above the Earth's surface. Calculate the speed at which the satellite is travelling.

a) 420 m s^{-1}
b) 490 m s^{-1}
c) 3100 m s^{-1}
d) 3300 m s^{-1}
e) 9.4×10^6 m s^{-1}

....

Q14: If the orbit radius of a satellite decreases,

a) its potential and kinetic energies both decrease.
b) its potential and kinetic energies both increase.
c) its potential energy increases but its kinetic energy decreases.
d) its potential energy decreases but its kinetic energy increases.
e) its potential energy decreases but its kinetic energy remains constant.

....

Q15: Astronomical observations of the period T and orbit radius r of a moon around a planet allow us to calculate the mass m_p of the planet. Which of these equations correctly expresses m_p in terms of T and r?

a) $m_p = 2\pi r^3 / CT^2$
b) $m_p = \sqrt{4\pi r^3 / GT^2}$
c) $m_p = 4\pi^2 r^2 / GT^3$
d) $m_p = 4\pi^2 r^3 / GT^2$
e) $m_p = 4\pi r / GT$

....

5.6   Gravitational potential and potential energy

We are now going to cover Gravitational potential and Gravitational potential energy.

5.6.1   Gravitational potential

The concept of gravitational potential energy should already be familiar to you from Newtonian mechanics. If a mass m is raised through a height h, it gains potential energy mgh. If the mass is then allowed to fall back down, this potential energy is converted to kinetic energy as it is accelerated downwards.

When an object moves through large distances in a gravitational field, we can no longer use this simple expression for the change in potential energy, as the value of the gravitational field strength g is not constant. The gravitational potential is used to describe how the potential energy of an object changes with its position in a gravitational field, and how much work is done in moving an object within the field.

The **gravitational potential** V at a point in a gravitational field is defined as the work done by external forces in moving a unit mass from infinity to that point. Suppose we are considering the field around a mass m, and moving a unit mass from infinity to a point a distance r from m. We cannot use the simple expression *work done = force× distance* to calculate the work done against the gravitational force, as the gravitational force acting on the unit mass increases as it moves closer to m. Instead, we have to use a calculus approach, in which we consider the small amount of work dV done in moving a unit mass a distance dr in the field. Integrating over the range from ∞ to r

$$V = \int_{\infty}^{r} F \, dr$$

Using Newton's law of gravitation for the force acting on the body

$$V = \int_{\infty}^{r} \frac{Gm_1m_2}{r^2} \, dr$$

In this expression, the values of m_1 and m_2 are M and 1 kg

$$V = \int_{\infty}^{r} \frac{GM}{r^2} \, dr$$

Performing this integration

$$V = GM \int_{\infty}^{r} \frac{1}{r^2}\, dr$$

$$\therefore V = GM \left[\frac{-1}{r} \right]_{\infty}^{r}$$

$$\therefore V = GM \left(-\frac{1}{r} - \left(-\frac{1}{\infty} \right) \right)$$

$$\therefore V = GM \left(\frac{1}{\infty} - \frac{1}{r} \right)$$

$$\therefore V = -\frac{GM}{r}$$

(5.10)

. . . .

The gravitational potential V is measured in J kg^{-1}. Remember that there is a minus sign in Equation 5.10. The zero of gravitational potential is at an infinite distance from M. The potential becomes lower and lower the closer we get to M, so it must be a negative number. The gravitational potential at a distance r from the Earth is plotted in Figure 5.5.

Figure 5.5: Gravitational potential around the Earth

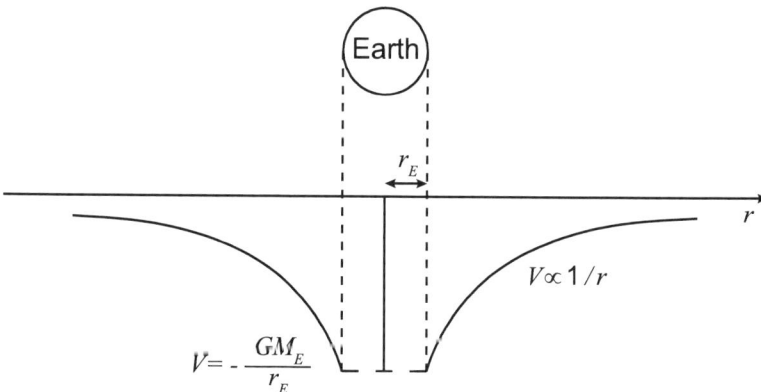

Example

The planet Pluto orbits at a mean distance of 5.92×10^{12} m from the Sun. What is the gravitational potential due to the Sun's gravitational field at this distance?

We will use Equation 5.10, remembering that M here represents the mass of the Sun, since we are calculating the potential due to the Sun's gravitational field. The mass of the Sun is 2.0×10^{30} kg, so

$$V = -\frac{GM}{r}$$
$$\therefore V = -\frac{6.67 \times 10^{-11} \times 2.0 \times 10^{30}}{5.92 \times 10^{12}}$$
$$\therefore V = -2.3 \times 10^7 \text{ J kg}^{-1}$$

The gravitational potential due to the Sun's gravitational field is -2.3 $\times$ 10^7 J kg^{-1}.

. . . .

5.6.2 Gravitational potential energy

Equation 5.10 tells us the gravitational potential at a point in a gravitational field per unit mass. To find the gravitational potential energy of an object of mass m_2 placed at that point in the field around a mass m_1, we simply multiply by m_2.

$$E_P = V \times m_2 = -\frac{Gm_1m_2}{r}$$

(5.11)

. . . .

We can now calculate the work done in moving an object in a gravitational field using this equation. The work done is equal to the change in potential energy of the object. When carrying out these calculations, it is important to take care to get the sign correct.

Example

A rocket ship, mass 4.00×10^5 kg, is travelling away from the Moon. The ship's rockets are fired when the ship is at a distance of 3.00×10^6 m from the centre of the Moon. If the mass of the Moon is 7.3×10^{22} kg, how much work is done by the rockets in moving the ship to a distance 3.20×10^6 m from the Moon's centre?

The rocket ship has an initial gravitational potential energy of

$$E_{P1} = -\frac{GM_Mm_s}{r}$$
$$\therefore E_{P1} = -\frac{6.67 \times 10^{-11} \times 7.3 \times 10^{22} \times 4.00 \times 10^5}{3.00 \times 10^6}$$
$$\therefore E_{P1} = -6.492 \times 10^{11} \text{ J}$$

The final value of the potential energy is

$$E_{P2} = -\frac{GM_M m_s}{r}$$
$$\therefore E_{P2} = -\frac{6.67 \times 10^{-11} \times 7.3 \times 10^{22} \times 4.00 \times 10^5}{3.20 \times 10^6}$$
$$\therefore E_{P2} = -6.086 \times 10^{11} \text{ J}$$

The change in potential energy ΔE_P is

$$\Delta E_P = E_{P2} - E_{P1}$$
$$\therefore \Delta E_P = -6.086 \times 10^{11} - \left(-6.492 \times 10^{11}\right)$$
$$\therefore \Delta E_P = 4.1 \times 10^{10} \text{ J}$$

The potential energy of the rocket ship has increased by 4.1×10^{10} J, so the amount of work done by the rockets must also be equal to 4.1×10^{10} J.

. . . .

The gravitational field is a **conservative field**, which means that the work done in moving a mass between two points in the field is independent of the path taken. In the above example, all that we are concerned with are the initial and final locations in the gravitational field. We do not need any details about the path taken between these two locations. (An example of doing non-conservative work would be the work done against friction in sliding a heavy mass between two points. Obviously we do less work if we slide the mass directly from one point to the other rather than taking a longer route.)

Quiz: Gravitational potential

Useful data:

Go online

Universal constant of gravitation G	6.67×10^{-11} m³ kg⁻¹ s⁻²
Mass of the Earth M_E	6.0×10^{24} kg
Radius of the Earth r_E	6.4×10^6 m
Mass of the Moon M_M	7.3×10^{22} kg
Radius of the Moon r_M	1.7×10^6 m

Q16: Consider the gravitational potential at a point A in the Earth's gravitational field. The value of the gravitational potential depends on

a) the mass of an object placed at A.
b) the speed of an object passing through A.
c) the distance of A from the centre of the Earth.
d) the density of an object placed at A.
e) the mass of the Earth only.

....

Q17: What is the gravitational potential on the surface of the Moon, due to the Moon's gravitational field?

a) -1.6 J kg^{-1}
b) -2.4×10^3 J kg^{-1}
c) -2.9×10^6 J kg^{-1}
d) -4.2×10^{17} J kg^{-1}
e) -2.9×10^{28} J kg^{-1}

....

Q18: A satellite orbits the Earth at an altitude of 3.00×10^5 m *above the Earth's surface*. What is the gravitational potential at this altitude?

a) -9.8 J kg^{-1}
b) -4.4×10^3 J kg^{-1}
c) -6.0×10^7 J kg^{-1}
d) -1.3×10^9 J kg^{-1}
e) -5.3×10^{24} J kg^{-1}

....

Q19: What is the gravitational potential energy of a satellite of mass 800 kg, moving in the Earth's gravitational field with orbit radius 6.60×10^6 m from the centre of the Earth?

a) -7.3×10^3 J
b) -6.1×10^7 J
c) -2.5×10^{10} J
d) -4.9×10^{10} J
e) -4.9×10^{32} J

....

Q20: The gravitational potential energy of a satellite orbiting the Earth changes from -5.0×10^9 J to -7.0×10^9 J. Which *one* of the following statements could be true?

a) The satellite has moved closer to the Earth.
b) The satellite has moved further away from the Earth.
c) The mass of the satellite has decreased only.
d) The orbit and mass have stayed the same, but the satellite is moving faster.
e) The orbit and mass have stayed the same, but the satellite is moving slower.

....

5.7   Escape velocity

At the beginning of the section on satellite motion, you should have carried out an interactive activity in which a projectile was launched into orbit with different speeds. If the speed was sufficiently great, the projectile did not complete an orbit, but escaped from the Earth's gravitational field.   We will now calculate what minimum speed is required for a projectile to escape from the Earth's gravitational field, and then generalise it to all other gravitational fields.

To escape from the gravitational field, an object must have sufficient kinetic energy. We know that at an infinite distance from Earth, the potential energy of a body will be zero. We can also work out the (negative) potential energy of the body when placed at the Earth's surface. If the body has sufficient kinetic energy to raise its total energy above zero J, then it can escape from the gravitational field.

The gravitational potential energy of an object such as a rocket of mass m at a point on the Earth's surface can be calculated using Equation 5.11.

$$E_P = -\frac{GM_E m}{r_E}$$

To escape the Earth's gravitational field, the work done by the rocket must equal the potential difference between infinity and the point on the Earth's surface.

$$\Delta E_P = \frac{-GM_E m}{\infty} - \frac{-GM_E m}{r_E}$$
$$\therefore \Delta E_P = 0 + \frac{GM_E m}{r_E}$$
$$\therefore \Delta E_P = \frac{GM_E m}{r_E}$$

The rocket must have an initial kinetic energy at least equal to this, so that its velocity does not drop to zero before it has escaped from the field. We can use the equation for kinetic energy to calculate the initial velocity of the rocket.

$$\frac{1}{2}mv^2 = \frac{GM_E m}{r_E}$$
$$\therefore v^2 = \frac{2GM_E}{r_E}$$
$$\therefore v = \sqrt{\frac{2GM_E}{r_E}}$$

(5.12)

· ·

This velocity is the minimum velocity required. Now we will use the expression for the acceleration due to gravity g, equivalent to the gravitational field strength at the Earth's surface, which is given by the equation

$$g = \frac{GM_E}{r_E{}^2}$$
$$\therefore g r_E = \frac{GM_E}{r_E}$$

Substituting this expression into the equation for the escape velocity gives us

$$v = \sqrt{\frac{2GM_E}{r_E}}$$
$$\therefore v = \sqrt{2g r_E}$$

(5.13)

· ·

The escape velocity for a rocket fired from Earth is given by Equation 5.13. Putting in the values of g = 9.8 m s^{-2} and r_E = 6.4 × 10^6 m, the escape velocity has a value

$$v = \sqrt{2g r_E}$$
$$\therefore v = \sqrt{2 \times 9.8 \times 6.4 \times 10^6}$$
$$\therefore v = 1.1 \times 10^4 \, \mathrm{m\,s^{-1}} \text{ or } 11 \, \mathrm{km\,s^{-1}}$$

You should note that the escape velocity does not depend on the mass of the rocket - the escape velocity is the same for any object launched from the Earth's surface. In general, the escape velocity from the gravitational field around a body of mass m, starting from a point r from the centre of the field, is given by the following equation

$$v = \sqrt{\frac{2GM}{r}}$$

(5.14)

....

From Equation 5.14, we can see that the escape velocity is greater for a planet with a higher mass. This causes interesting differences between the planets' atmospheres. The Earth's atmosphere is mainly nitrogen and oxygen, with a very low incidence of helium. In contrast, larger planets such as Jupiter have plenty of helium. Molecules such as helium have a very low mass. This meant that when the Earth was forming, the helium molecules were able to reach velocities in excess of the Earth's escape velocity and they managed to escape into space. Meanwhile, Mercury has an extremely thin atmosphere made up of atoms blasted off its surface by the solar wind (Unit 2 - Topic 2). Mercury's atmosphere is constantly escaping and being replaced, since its high temperature means that its gas molecules can travel faster than the comparatively low escape velocity.

Example

What is the escape velocity for a lunar probe taking off from the surface of the Moon? (M_M = 7.3 × 10^{22} kg, r_M = 1.7 × 10^6 m)

Using Equation 5.14 with the values given in the question, the escape velocity from the Moon is

$$v = \sqrt{\frac{2GM_M}{r_M}}$$

$$v = \sqrt{\frac{2 \times 6.67 \times 10^{-11} \times 7.3 \times 10^{22}}{1.7 \times 10^6}}$$

$$v = \sqrt{5.728 \times 10^6}$$

$$v = 2.4 \times 10^3 \, \text{m s}^{-1}$$

The escape velocity from the Moon's gravitational field is 2.4 × 10^3 m s^{-1}, or 2.4 km s^{-1}.

....

Quiz: Escape velocity

Useful data:

Go online

Universal constant of gravitation G	6.67 x 10^{-11} m^3 kg^{-1} s^{-2}
Mass of the Earth M_E	6.0 × 10^{24} kg
Radius of the Earth r_E	6.4 × 10^6 m

Q21: The escape velocity of an object taking off from Earth is the minimum velocity required to

a) place the object in a geostationary orbit.
b) place the object in a non-geostationary orbit.
c) reach the point where the combined field of the Earth and the Moon is zero.
d) escape from the Earth's atmosphere.
e) escape from the Earth's gravitational field.

. . . .

Q22: The escape velocity of an object from the Earth depends on

a) the masses of the Earth and the object.
b) the mass and radius of the Earth.
c) the mass and density of the object.
d) the mass of the object and the radius of the Earth.
e) the mass and radius of the Earth, and the mass of the object.

. . . .

Q23: What would be the escape velocity of a Martian spacecraft of mass 900 kg taking-off from the surface of Mars, if the mass and radius of Mars are 6.42×10^{23} kg and 3.40×10^{6} m?

a) 2510 m s^{-1}
b) 3550 m s^{-1}
c) 5020 m s^{-1}
d) $1.06 \times 10^{5} \text{ m s}^{-1}$
e) $1.51 \times 10^{5} \text{ m s}^{-1}$

. . . .

5.8   Extended information

Web links

There are web links available online exploring the subject further.

. . . .

5.9 Summary

---Summary---

You should now be able to:

- state and apply the equation $F \frac{GMm}{r^2}$ to calculate the gravitational force between two objects;

- calculate the weight of an object using Newton's Universal Law of Gravitation;

- state what is meant by a gravitational field, and calculate the gravitational field strength at a point in the field;

- calculate the value of the acceleration due to gravity at a point in a gravitational field, given the local conditions;

- sketch the field lines around a planet and a planet-moon system;

- use appropriate relationships to carry out calculations involving the period of a satellite in circular orbit, a satellite's mass, speed and orbital radius;

- explain what is meant by a geostationary satellite;

- state the expression $V = \frac{-GM}{r}$ and use it to calculate the gravitational potential V at a point in a gravitational field;

- calculate the gravitational potential energy of a mass in a gravitational field and calculate the change in the potential energy when a mass is moved between points in the field;

- define escape velocity as the minimum velocity required to allow a mass to escape a gravitational field;

- derive the expression $v = \sqrt{\frac{2GM}{r}}$ and use it to calculate the escape velocity.

5.10 Assessment

End of topic 5 test

The following test contains questions covering the work from this topic.

Go online

The following data should be used when required:

Universal constant of gravitation G	$6.67 \times 10^{-11}\ m^3\ kg^{-1}\ s^{-2}$
Mass of the Earth M_E	$6.0 \times 10^{24}\ kg$
Mass of the Moon M_M	$7.3 \times 10^{22}\ kg$
Mass of the Sun M_S	$2.0 \times 10^{30}\ kg$
Radius of the Earth r_E	$6.4 \times 10^6\ m$
Radius of the Moon r_M	$1.7 \times 10^6\ m$

Q24: A distant planet has mass 5.45×10^{25} kg. A moon, mass 3.04×10^{22} kg, orbits this planet with an orbit radius of 7.16×10^8 m.

Calculate the size of the gravitational force that exists between the moon and the planet.

F = _____ N

....

Q25: Two identical solid spheres each have mass 0.853 kg and diameter 0.245 m.

Find the gravitational force between them when they are touching.

F = _____ N

....

Q26: The mass of planet Neptune is 1.03×10^{26} kg and its radius is 2.48×10^7 m.

Calculate the weight of a 6.64 kg mass on the surface of Neptune.

Weight on Neptune = _____ N

....

Q27: The value of the acceleration due to gravity is not constant, decreasing with height above the surface of the Earth.

What is the value of the acceleration due to gravity in the ionosphere at a height 3.03×10^5 m above the Earth's surface?

g = _____ m s^{-2}

....

Q28: On the surface of the Earth, a particular object has a weight of 22.0 N.

Calculate its weight on the surface of the Moon.

_____ N

....

Q29: A planet in a distant galaxy has mass 6.67×10^{25} kg and radius 4.02×10^7 m.

Calculate the value of the gravitational field strength on the surface of this planet.

_____ N kg^{-1}

....

Q30: The gravitational field strength at a distance 2.13×10^6 m from the centre of a planet is 6.05 N kg^{-1}.

Calculate the field strength at a distance 8.52×10^6 m from the centre of the planet.

Note: It is possible to solve this problem without having to calculate the mass of the planet.

_____ N kg^{-1}

. . . .

Q31: In a distant solar system, a planet (mass 2.02×10^{28} kg) is orbiting a star (mass 5.41×10^{30} kg) with an orbit radius of 4.44×10^{11} m.

Calculate the magnitude of the net gravitational field strength midway between the planet and the star.

_____ N kg^{-1}

. . . .

Q32: Calculate the gravitational potential at a distance 1.25×10^7 m from the centre of the Earth, due to the Earth's gravitational field.

$V =$ _____ J kg -1

. . . .

Q33: A spaceship, mass 5.4×10^4 kg is travelling through the solar system. At one point in its journey, the spaceship passes near the planet Jupiter taking photographs at a distance 9.5×10^7 m from the centre of Jupiter.

If the mass of Jupiter is 1.9×10^{27} kg, calculate the potential energy of the spaceship at this point.

$E_p =$ _____ J

. . . .

Q34: The gravitational potential at a distance 2.46×10^7 m from the centre of a planet of radius 8.55×10^6 m is -5.24×10^7 J kg^{-1}.

Calculate the gravitational potential at a distance 4.92×10^7 m (twice the original distance) from the centre of the planet.

Note: It is possible to solve this problem without having to calculate the mass of the planet.

$V =$ _____

. . . .

Q35: A meteorological satellite is orbiting the Earth. The mass of the satellite is 5.25×10^3 kg and it orbits at a height 1.35×10^5 m above the Earth's surface.

Calculate the gravitational potential energy of the satellite.

$E_p =$ _____ J

. . . .

Q36: Scientists wish to launch a satellite (mass 2.8×10^4 kg) which will orbit the Earth once every 5400 seconds.

At what height above the Earth's surface should the satellite be placed?

_____ m

. . . .

Q37: The planet Zaarg has mass 7.54×10^{25} kg and radius 2.88×10^7 m.

Calculate the escape velocity of a rocket ship launched from the planet Zaarg.

_____ m s^{-1}

. . . .

Q38: Astronomers observing a distant solar system have noticed a planet orbiting a star with a period 6.82×10^7 s. The distance from the planet to the star is 2.95×10^{11} m.

Calculate the mass of the star.

_____ kg

. . . .

Topic 6

General relativity and spacetime

Contents

Prerequisite knowledge

- *Special relativity (CfE Higher).*

- *Gravitation (Advanced Higher Unit 1 - Topic 5).*

Learning objectives

By the end of this topic you should be able to:

- *state that an inertial frame of reference is one that is stationary or has a constant velocity;*

- *state that a non-inertial frame of reference is one that is accelerating;*

- *state that Einstein's theory of special relativity is appropriate for inertial frames of reference;*

- *state that Einstein's theory of general relativity is appropriate for non-inertial frames of reference;*

- *describe Einstein's equivalence principle in terms of an accelerated frame of reference being equivalent to a reference frame at rest in a gravitational field;*

- *state that when an object is in freefall, the downwards acceleration exactly cancels out the effects of being in a gravitational field;*

- *explain some consequences of the equivalence principle, such as that clocks in a weaker gravitational field run faster than those in a stronger gravitational field;*

- *explain some of the pieces of evidence for general relativity;*

- *explain that a particle's worldline on a spacetime diagram shows its spatial location at every instant in time;*

- *interpret spacetime diagrams for stationary objects, those moving at a constant speed and accelerating objects;*

- *state that the greater the gradient of a worldline, the smaller the velocity;*

- *explain that curved lines on spacetime diagrams correspond to non-inertial (accelerating) frames of reference i.e. accelerations are represented by worldlines of changing gradient;*

- *identify simultaneous events on a spacetime diagram;*

- *explain that what we perceive as the force of gravity in fact arises from an object with large mass distorting spacetime;*

- *state that the density of a black hole is so great that its escape velocity is greater than the speed of light;*

- *explain what is meant by the Schwarzschild radius of a black hole;*

- *solve problems using the equation for the Schwarzschild radius of a black hole $r_{Schwarzschild} = \frac{2GM}{c^2}$;*

- *state that the event horizon is the boundary of a black hole and that no matter or radiation can escape from within the event horizon.*

6.1 Introduction

Einstein's special theory of relativity only applies when an object is moving at a constant speed in a straight line. It does not let us consider what happens when something turns or changes speed. For this we need Einstein's general theory of relativity.

In this topic we will compare the two theories and explore how **general relativity** allows us to more fully understand gravity. We will then turn our attention to the effect of mass on **spacetime** and the curious behaviour of black holes.

6.2 Comparison of special and general relativity

You studied the special theory of relativity at Higher. You may remember that in this theory Einstein made the following two points:

1. The speed of light is absolute. It is always the same for all observers irrespective of their relative velocities.

2. The laws of Physics are the same for all observers inside their frame of reference.

These observations lead to some conclusions that at first seemed strange but were confirmed by observation, such as time dilation and length contraction. However, the theory did not allow us to consider what happens when a frame of reference is accelerating. For this we will need general relativity.

In other words, **special relativity** only considered inertial frames of reference. That is ones that are stationary or have a constant velocity. General relativity will allow us to study a **non-inertial frame of reference**, which is one that is accelerating.

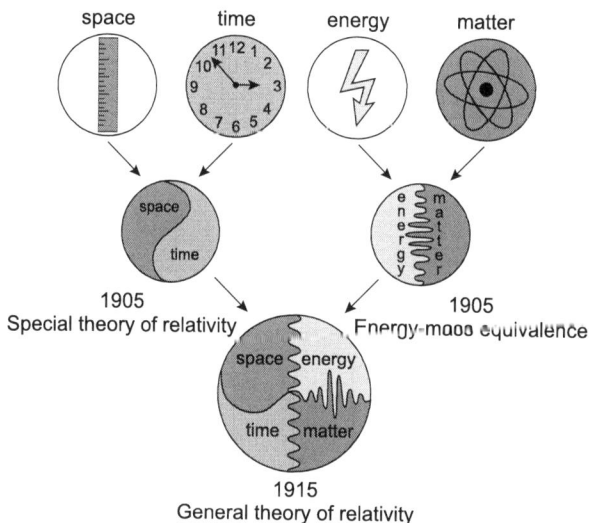

space   time   energy   matter

1905
Special theory of relativity

1905
Energy-mass equivalence

1915
General theory of relativity

6.3 The equivalence principle

Einstein used the following thought experiment to show how he was extending relativity to include accelerating objects. Consider two observers in identical spaceships, one at rest on the Earth and the other accelerating at 9.8 ms^{-2} in deep space, far away from any astronomical bodies. Einstein said it would be impossible to distinguish between the two situations and that an experiment would produce the same results in both spaceships. For instance, if both observers stand on a newton balance, then provided they have the same mass as each other, the readings would be identical in the two scenarios.

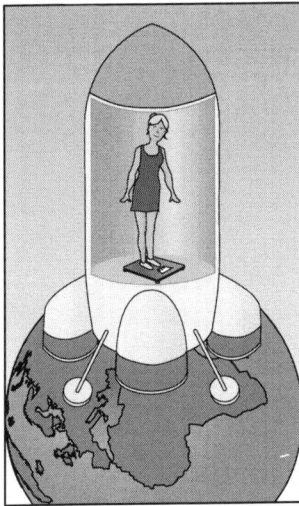   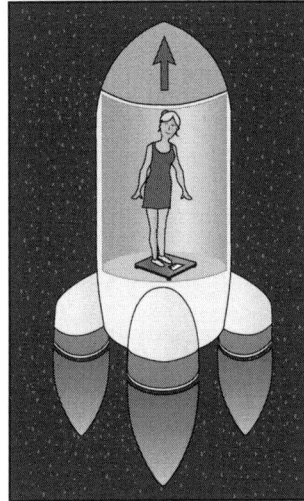

(a)   (b)

(a) At rest in the Earth's gravitational field. (b) In deep space accelerating upwards at 9.8 m s^{-2}.

Furthermore, a ball dropped in one spaceship would fall to the floor in the same way as in the other spaceship. When the ball is released in the spaceship on Earth, the gravitational field will cause it to accelerate downwards towards the floor at 9.8 ms^{-2}. When the ball is released in the spaceship in outer space, it becomes a free object. That is, there is no unbalanced force acting upon the ball. However, since the spaceship continues to accelerate upwards, the floor of the spaceship will accelerate upwards to meet the ball. So, from the observer's viewpoint, the ball appears to accelerate downwards towards the floor at 9.8 ms^{-2}.

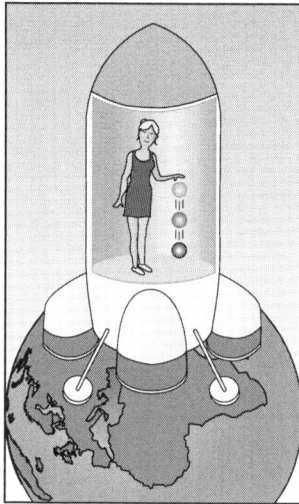 

(a)   (b)

*(a) At rest in the Earth's gravitational field. (b) In deep space accelerating upwards at 9.8 m s⁻². *

Einstein's **equivalence principle** summarises this. It states that it is impossible to tell the difference between a uniform gravitational field and a frame of reference that has a constant acceleration. In other words, gravity is equivalent to acceleration.

Einstein also pondered what would happen if a person dropped a ball whilst they were falling off the side of a building. He realised that a person who accelerates downwards along with the ball will not be able to detect the effects of gravity on the ball. Indeed, both the ball and the person would effectively be weightless in this scenario. They would be equivalent to a person motionless in a spaceship in deep space. He identified that a force is experienced when accelerating or when in a gravitational field, but no force is felt when an object is in freefall, since the downwards acceleration exactly cancels out the effect of being in a gravitational field.

So, all of these thought experiments made Einstein realise that the force of gravity is just the acceleration that you feel as you move through spacetime (see section 6.8).

The equivalence principle

There is an animation available online showing the effects of gravity.

. . . .

Go online

6.4   Understanding the consequences

Bending light

Consider a laser beam sent across a spaceship that is accelerating upwards. To an observer who watches the spaceship from the ground, the light moves in a straight line. However, the spaceship will move slightly upwards in the time it takes the light to travel across it. Therefore the light will strike a point lower on the spaceship wall than it would if the spaceship did not move. This means that, to an observer inside the spaceship, the light actually appears to bend.

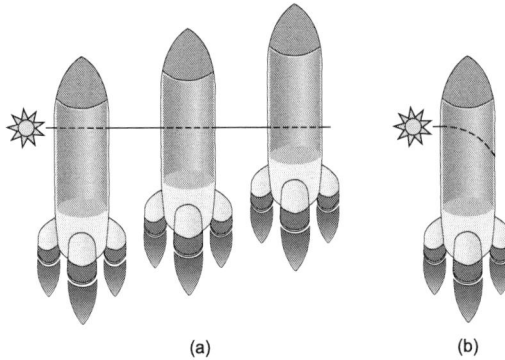

(a)   (b)

(a) View from the ground. (b) View inside the spaceship.

In the last section we explored the **equivalence principle**, which states that an accelerating frame acts in the same way as a frame in a gravitational field. Therefore, we can conclude that gravitational fields must also bend light. Since we are very familiar with light being bent by a lens, we call this effect gravitational lensing.

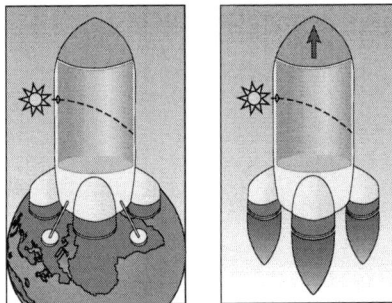

The Equivalence Principle - the bending of light due to acceleration and gravity.

Gravitational time dilation

Once more, consider a spaceship which is accelerating upwards. A source of light emits pulses at regular time intervals from the bottom of the spaceship. By the time a pulse reaches an observer at the top of the spaceship, the spaceship will have moved away from the position it occupied when the pulse was emitted. Since the spaceship is accelerating, the distance each consecutive light pulse must travel to reach the observer will be increasing. That means the time for pulses to reach the observer will also be increasing. So the observer at the top of the spaceship will conclude that the clock at the bottom of the spaceship is running more slowly. We already know from the equivalence principle that a spaceship accelerating upwards is equivalent to a spaceship at rest in a gravitational field.

The clock at the rear runs more slowly.

So for a clock in a gravitational field, time runs more slowly than in the absence of a field.

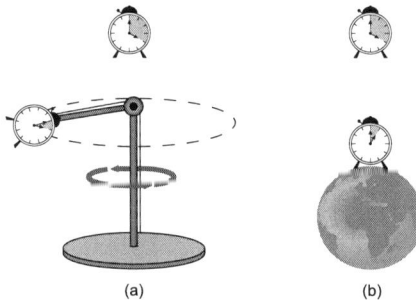

(a)   (b)

(a) Accelerating clock in a centrifuge. (b) Clock in a gravitational field - the lower the slower.

6.5   Show me the evidence

Gravitational red shift

Consider a source of light at the top of an accelerating spaceship. Due to the Doppler Effect, an observer at the bottom of the spaceship will observe a higher frequency than the source emitted. The **equivalence principle** means that a beam of light travelling in the same direction as a gravitational field should shift towards the higher frequency blue end of the spectrum. Furthermore, a beam of light travelling upwards from the Earth's surface should be shifted towards the lower frequency red end of the spectrum. More generally, any electromagnetic wave originating from a source that is in a gravitational field is reduced in frequency when observed in a region of a weaker gravitational field. We call this effect gravitational red shift. It was confirmed experimentally for the first time by measuring the relative redshift of two sources situated at the top and bottom of a tower at Harvard University.

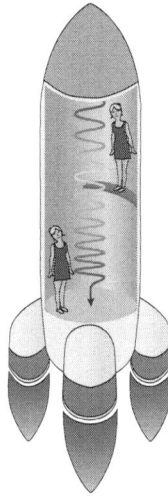

This effect can also be explained in terms of the photons. If a photon moves towards the Earth, the photon's energy increases as it travels to a position of smaller gravitational potential energy. Therefore, from $E = hf$, the photon's frequency ought to increase.

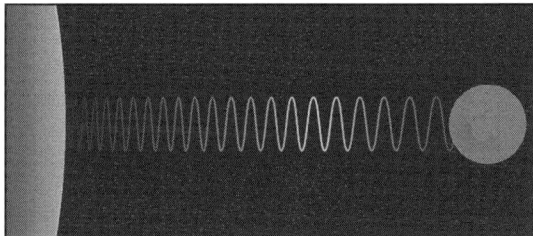

GPS

When Einstein first produced his general theory of relativity, its predictions were far removed from everyday experience and were therefore difficult to test experimentally. However, most people are nowadays accustomed to relying on the theory to keep their GPS (Global Positioning System) receivers working accurately.

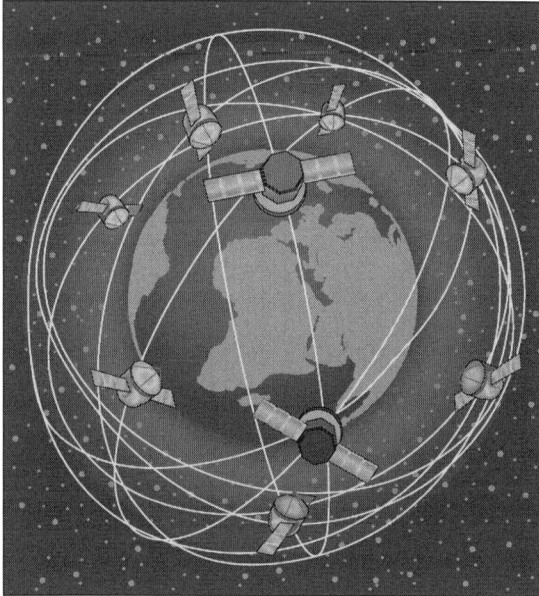

The clocks in GPS satellites need to be adjusted to take account of relativity.

Since the Earth's gravitational field is weaker at altitude, a clock in a GPS satellite will run fast relative to a clock on the ground. Adjustments need to be made to take this into account or the clocks will become out of sync within minutes. Corrections also need to be made to allow for the special relativistic effect of time dilation brought about by the large velocity of the satellite. Ignoring Einstein's theories would result in a discrepancy of around 10 km per day. So within a week you would be confusing Edinburgh with Glasgow.

6.6 Spacetime diagrams

In the last topic we saw that space and time are linked and that gravity is a property of both space and time. So it is helpful to consider spacetime, which is a coordinate system that involves the three dimensions of space (x, y and z), along with time. The motion of a particle can then be plotted as a series of points in this system.

Consider an event (a particular place at a particular time). Let's call it E. Imagine light moving out from a source at E in an expanding spherical shell. To simplify matters, let's consider only two dimensional space for just now. That is, let's consider x and y, but ignore z. Then the light would look like an expanding circle. Now imagine snapshots are taken at regular intervals and stacked on top of each other so that the horizontal plane represents how far the light has moved and the vertical axis represents time. A cone shape would be formed.

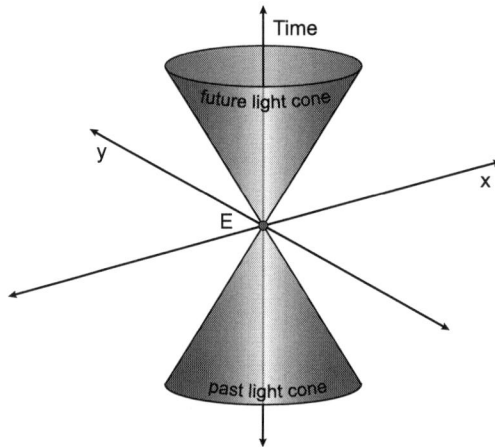

We constructed this lightcone by choosing an event in spacetime and imagining all the possible paths that light could take in moving through this event. However, we could have considered any event in spacetime and its corresponding lightcone. So spacetime is actually filled with light cones; there is one for every event.

Knowing the light cone structure of spacetime allows us to see which events can have an effect upon others. The top half of the light cone represents the future and it is the all events that have not yet happened that could be affected by event E. The bottom half is the past and it is all the events that could have contributed to event E. A region outside the light cone would be inaccessible, since something would need to travel faster than light to reach it. Remember from special relativity that the speed of light is believed to be the greatest possible speed.

To understand this, let us consider Alpha Centauri, our next nearest star after the Sun. It is 4.4 light years from Earth. Now, if you consider yourself at this present moment in time as an event, then Alpha Centauri in three years' time would lie outside your light cone. In other words, you cannot reach it without travelling faster than the speed of light. Therefore, it is not accessible. However, Alpha Centauri in thirty years' time does lie inside your light cone. This means that theoretically you could travel there, though we do not currently have the technology to reach Alpha Centauri within a human's lifetime.

The path of an object through spacetime is called a worldline. An example is shown in red below. Note that wordlines cannot go beyond the light cone, as to do so would require an object to travel at a speed greater than the speed of light.

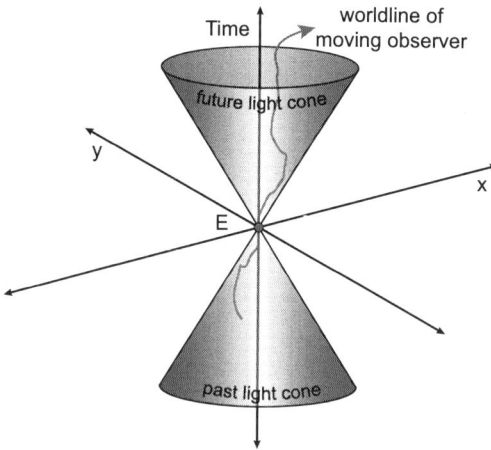

In reality, the light cone would actually be four-dimensional, 3 dimensions for space and one for time. However, the concept is easier to visualise with the number of spatial dimensions reduced. Indeed, the diagrams are most often simplified to show time and only one spatial dimension as shown below.

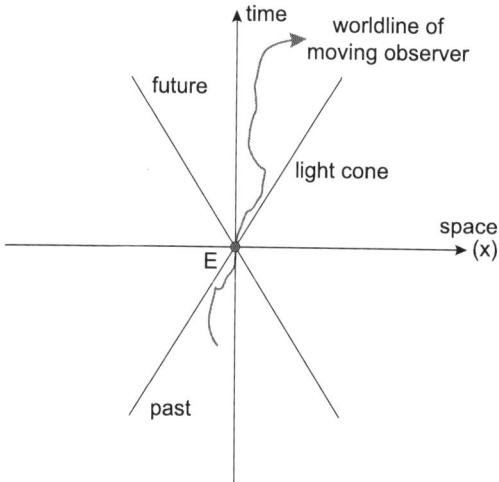

Again, the different regions of the spacetime diagram have the following significance:

- The diagonal lines above correspond to a speed equal to the speed of light in vacuum i.e. $v = c$.

- The regions within the diagonals are where the speed is less than c.

- The regions outside the diagonals cannot affect or be affected by the event at E, since messages cannot travel faster than c.

- The present corresponds to where the time is zero.

- The region below the x-axis is the past.

Spacetime diagram

There is an animation available online displaying the wordlines for four objects moving through spacetime.

Go online

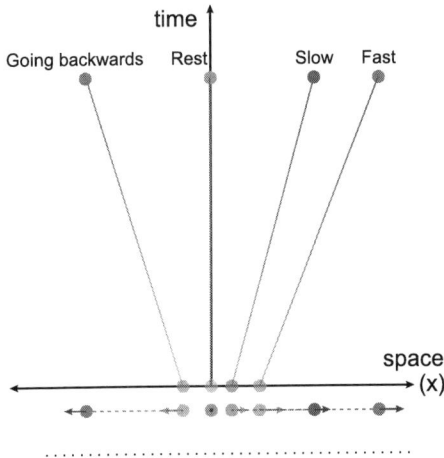

6.7 Worldlines

As you found out in the last section, a line on a spacetime diagram which maps a particle's spatial location at every instant in time is called a **worldline**. Typically only events in a one dimensional world are considered as this is a lot simpler to understand! Each point of a worldline is an event that can be labelled with the time and the spatial position of the object at that particular time. Note that unlike distance-time graphs, the spatial position will be displayed on the x-axis and the time on the y-axis.

A stationary object's position does not change with time and so its spacetime diagram would be as follows.

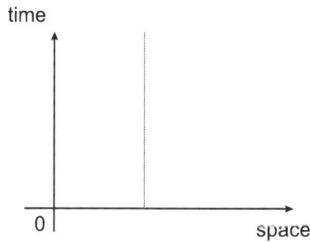

An object moving at constant speed could be shown by either of the following worldines.

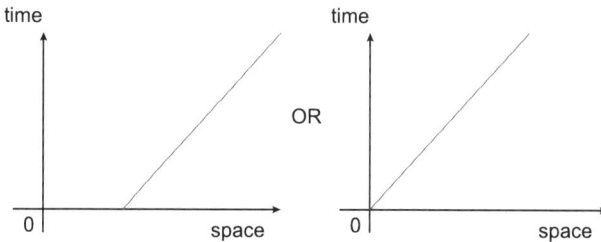

OR

The greater the gradient of a worldline, the smaller the velocity.

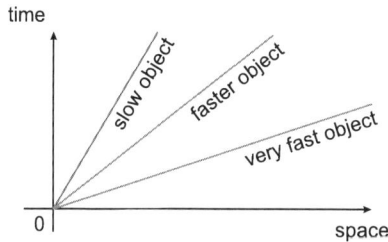

Accelerating objects are shown on spacetime diagrams as having world lines of changing gradient. So curved lines on spacetime diagrams correspond to non-inertial frames of reference i.e. accelerating frames of reference.

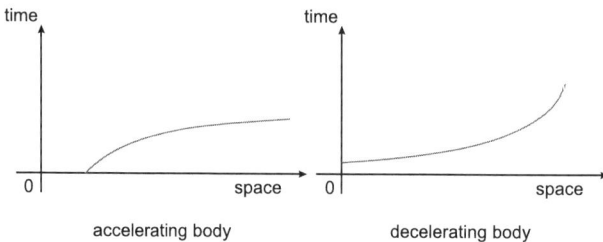

accelerating body   decelerating body

Simultaneous events are ones that occur at the same time. Therefore simultaneous events are shown as a flat line on a spacetime diagram, a line of constant time.

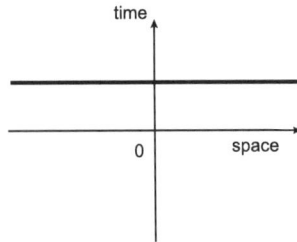

6.8   The curvature of spacetime

We saw earlier that gravitational lensing is the process whereby a large mass can make light bend. In this situation, the light is infact still travelling in a straight line. It is just that the mass has actually warped spacetime into being curved. Light always travels the shortest path between two points in spacetime and this path is called a **geodesic path**.

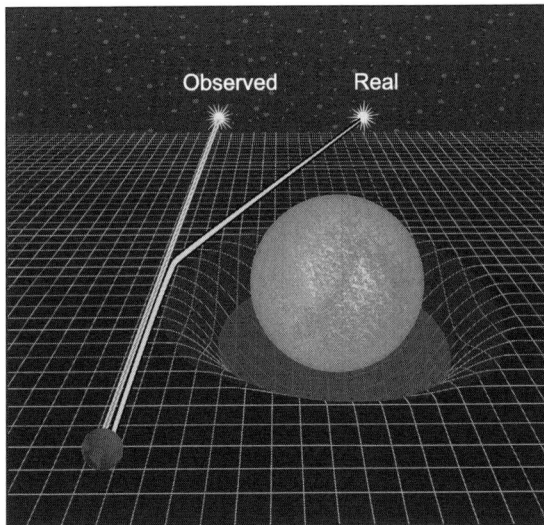

Warped spacetime

Einstein proposed that a large mass curves and stretches the spacetime around it, rather like a bowling ball creating a dimple as it distorts a rubber sheet. This then affects the motion of any other body that enters its spacetime since it will now need to follow the curvature to follow a straight line. So what is called gravity is really the result of the mass of an object creating a curvature in spacetime. Just as an object with a greater mass causes a greater distortion in a rubber sheet, an astronomical body with a larger mass causes a greater distortion and curvature in the fabric of spacetime. So gravity feels strongest where spacetime is most curved, and it vanishes where spacetime is flat.

The same explanation can be used to account for the motion of an object in freefall, such as a satellite, where the only force acting on it is gravity. The gravitational field of the central body makes spacetime curved so that the satellite's geodesic in spacetime is now curved, rather than straight.

Warped spacetime

There is an online activity showing animations of the warped spacetime.

Go online

   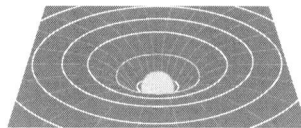

(a)   (b)

Spacetime is flat without matter (a), but it curves when matter is present (b).

....

In short, matter tells spacetime how to curve, and curved spacetime tells matter how to move.

The gravitational force of attraction between the Sun and a planet decreases with distance from the Sun. The closer to the Sun, the greater the degree of curvature in spacetime. The further from the Sun, the less the curvature in spacetime.

The mass of the Sun causes spacetime to curve, so each planet follows the shortest and straightest possible path allowed by the curvature of spacetime.

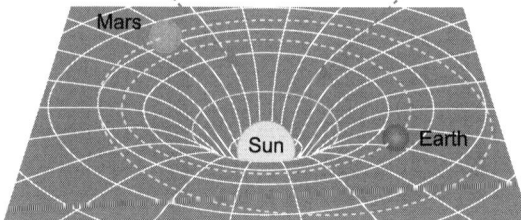

Gravity Probe B was launched in 2004 to measure the curvature of spacetime. Tiny deviations in the orientation of spinning gyroscopes allowed astronomers to measure the amount by which the Earth warps the local spacetime in which it resides.

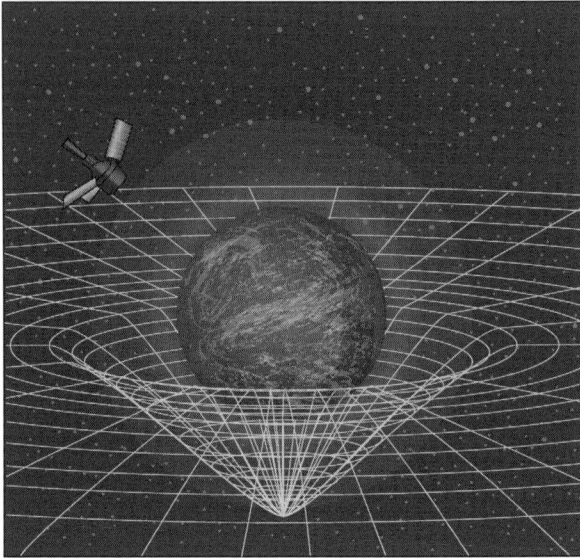

The orbit of Mercury

Einstein's general theory of relativity has now been demonstrated experimentally in various ways. However, one of the earliest pieces of evidence to support it was the motion of Mercury. Newtonian Physics allowed astronomers to accurately describe the orbit of the planets in our solar system, but they remained mystified by the behaviour of Mercury. They understood that the perturbing (changing) effect of the other planets made Mercury's elliptical orbit **precess** around the Sun, like a spinning top. However, Newtonian Physics could not account for the extent to which Mercury precessed. Einstein managed to successfully explain Mercury's behaviour by outlining that the mass of the Sun was creating a curvature in spacetime. In other words, the Sun was warping spacetime and Mercury was simply following the resulting curvature in the fabric of spacetime.

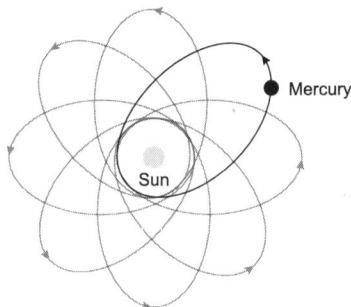

Observing gravitational lensing

General relativity predicts that an astronomical body with a very large mass ought to bend light. The astronomer Arthur Eddington was the first to provide evidence to confirm this. He made measurements to show that the gravitational field of the Sun deflects light from its straight path towards the Earth. Furthermore, **gravitational lensing** by various astronomical bodies has now been verified numerous times with data from radio telescopes.

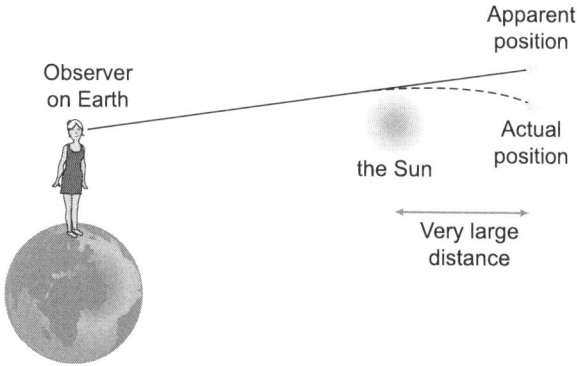

An interesting example of this effect is known as an Einstein ring. If the light from a distant galaxy is made to bend by the gravitational field of another galaxy that is directly behind it, the light can be focused into a visible ring.

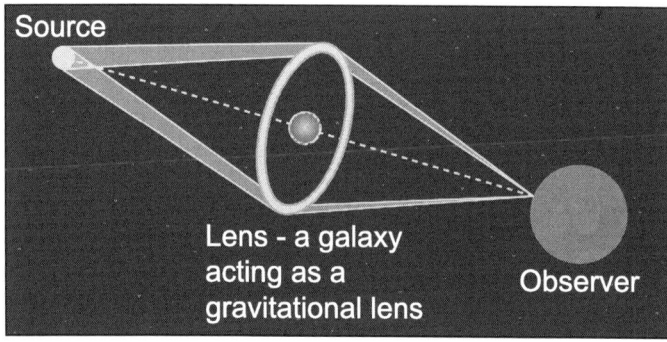

Gravitational waves - Ripples in spacetime

Einstein's general theory of relativity also predicts the existence of **gravitational waves**, which are ripples in spacetime generated during extremely violent astrophysical events, such as the collision of two black holes.

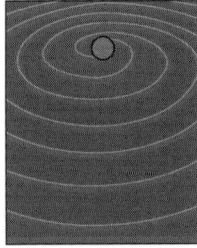

Gravitational waves are so weak that they are very hard to detect, but special interferometers such as LIGO (Laser Interferometer Gravitational Wave Observatory) use precisely calibrated laser beams to search for them. A passing gravitational wave ought to slightly distort spacetime and cause a noticeable shift in the interference pattern created by the lasers.

Image provided by Caltech/MIT/LIGO Laboratory

Astronomers' observations are currently limited to objects which emit electromagnetic radiation, but vast parts of the Universe are obscured by dark clouds. Since gravitational waves can pass through unhindered, their detection would allow astronomers to greatly increase their knowledge. Furthermore, the Big Bang is believed to have created a flood of gravitational waves which still fill the Universe today. So the detection of gravitational waves would also allow astronomers to gain a better understanding of the creation, development and fate of the Universe. It is not surprising some people call them Einstein's messengers.

6.9   Black holes

We will see in Topic 7 that when a star of exceptionally large mass reaches the end of its life, gravitational compression will cause it to collapse to a very small radius, producing an incredibly dense body called a **black hole**. Due to their extraordinary density, black holes exert extremely strong gravitational fields. Or rather, as general relativity would describe it, black holes severely distort spacetime.

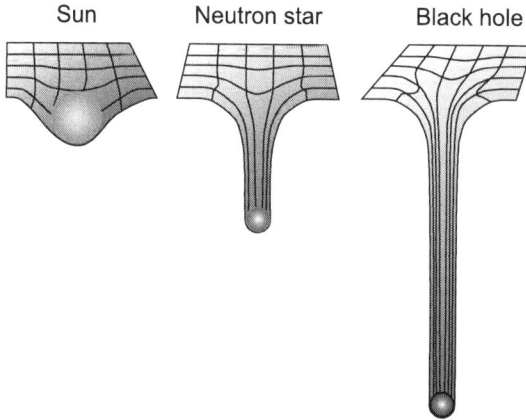

A black hole's extremely large density means it severely distorts the surrounding spacetime.

The spacetime around a black hole is stretched to a point of infinite density known as a **singularity**. This is a single point to which all mass would collapse.

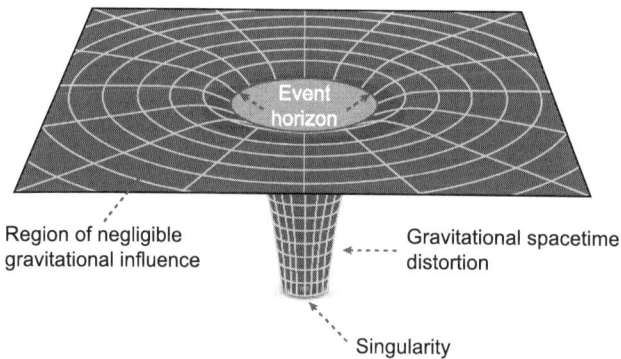

Black hole regions

Up to a certain distance away (**Schwarzschild radius**), the gravitational field around a black hole is so high that nothing can escape, not even light. Another way to look at it is that for distances less than the Schwarzschild radius, the escape velocity is greater than the speed of light. And since nothing can travel faster than the speed of light, nothing can escape, not even photons. This means that a black hole's name is certainly a very apt description. A black hole is effectively cut off from the rest of the Universe. However, a black hole's presence can still be deduced by detecting the stream of X-rays produced as matter falls into it.

The Schwarzschild radius and the event horizon

The Schwarzschild radius is the distance from the centre of a black hole at which not even light can escape. At the Schwarzschild radius the escape velocity is equal to the speed of light.

In Topic 5 we saw that the escape velocity for an object in a gravitational field can be found from:

$$v = \sqrt{\frac{2GM}{r}}$$

So, replacing v with the speed of light, c, gives the equation for the Schwarzschild radius:

$$c = \sqrt{\frac{2GM}{r_{Schwarzschild}}}$$

Squaring both sides results in:

$$c^2 = \frac{2GM}{r_{Schwarzschild}}$$

Rearranging for $r_{Schwarzschild}$ gives:

$$r_{Schwarzschild} = \frac{2GM}{c^2}$$

The Schwarzschild radius is also called the gravitational radius. It effectively forms a boundary called the **event horizon**. No matter or radiation can escape from within the event horizon. It is possible to go from outside the event horizon to inside, but once you are over it, there is no going back!

Launching a clock into a black hole

There is an online activity showing an animation of a clock launched into a black hole.

Go online

. . . .

Examples

1. A star of mass 7.96×10^{32} kg collapses to form a black hole. Calculate the Schwarzschild radius.

$$r_{Schwarzschild} = \frac{2GM}{c^2}$$
$$r_{Schwarzschild} = \frac{2 \times 6.67 \times 10^{-11} \times 7.96 \times 10^{32}}{\left(3 \times 10^8\right)^2}$$
$$r_{Schwarzschild} = 1.18 \times 10^6 m$$

. . . .

2. Calculate the mass of a black hole with a Schwarzschild radius of 1.76×10^3 km.

$$r_{Schwarzschild} = \frac{2GM}{c^2}$$
$$1.76 \times 10^6 = \frac{2 \times 6.67 \times 10^{-11} \times M}{\left(3 \times 10^8\right)^2}$$
$$M = 1.19 \times 10^{33} \text{kg}$$

. . . .

3. A black hole has a mass of 1.50×10^{34} kg. Find the distance of the event horizon from its centre.

$$r_{Schwarzschild} = \frac{2GM}{c^2}$$
$$r_{Schwarzschild} = \frac{2 \times 6.67 \times 10^{-11} \times 1.50 \times 10^{34}}{\left(3 \times 10^8\right)^2}$$
$$r_{Schwarzschild} = 2.22 \times 10^7 m$$

. . . .

4. Calculate the mass of a black hole with a Schwarzschild radius of 2.98×10^{29} km.

$$r_{Schwarzschild} = \frac{2GM}{c^2}$$

$$2.98 \times 10^{32} = \frac{2 \times 6.67 \times 10^{-11} \times M}{(3 \times 10^8)^2}$$

$$M = 2.01 \times 10^{59} \text{kg}$$

....

6.10 Summary

Summary

You should now be able to:

- state that an inertial frame of reference is one that is stationary or has a constant velocity;

- state that a non-inertial frame of reference is one that is accelerating;

- state that Einstein's theory of special relativity is appropriate for inertial frames of reference;

- state that Einstein's theory of general relativity is appropriate for non-inertial frames of reference;

- describe Einstein's equivalence principle in terms of an accelerated frame of reference being equivalent to a reference frame at rest in a gravitational field;

- state that when an object is in freefall, the downwards acceleration exactly cancels out the effects of being in a gravitational field;

- explain some consequences of the equivalence principle, such as that clocks in a weaker gravitational field run faster than those in a stronger gravitational field;

- explain some of the pieces of evidence for general relativity;

- explain that a particle's worldline on a spacetime diagram shows its spatial location at every instant in time;

- interpret spacetime diagrams for stationary objects, those moving at a constant speed and accelerating objects;

- state that the greater the gradient of a worldline, the smaller the velocity;

- explain that curved lines on spacetime diagrams correspond to non-inertial (accelerating) frames of reference i.e. accelerations are represented by worldlines of changing gradient;

Summary continued

- identify simultaneous events on a spacetime diagram;

- explain that what we perceive as the force of gravity in fact arises from an object with large mass distorting spacetime;

- state that the density of a black hole is so great that its escape velocity is greater than the speed of light;

- explain what is meant by the Schwarzschild radius of a black hole;

- solve problems using the equation for the Schwarzschild radius of a black hole $r_{Schwarzschild} = \frac{2GM}{c^2}$;

- state that the event horizon is the boundary of a black hole and that no matter or radiation can escape from within the event horizon.

6.11 Extended information

Web links

There are web links available online exploring the subject further.

. . . .

6.12 Assessment

End of topic 6 test

The following test contains questions covering the work from this topic.

Go online

The following data should be used when required:

Gravitational acceleration on Earth g	$9.8\ m\ s^{-2}$
Gravitational acceleration on Moon g	$1.6\ m\ s^{-2}$

Q1: Einstein's theory of general relativity considers frames of reference that are

---------- .

a) inertial
b) non-inertial

. . . .

Q2: It deals with objects that are _____ .

a) stationary
b) moving at constant velocity
c) accelerating

....

Q3: Einstein identified that there is no way of distinguishing between the effects on an observer of a uniform gravitational field and of constant acceleration.
This is called his _____ .

....

Q4: In a gravitational field time runs more _____ .

a) slowly
b) quickly

....

Q5: When at the rear of an accelerating object, time passes _____ .

a) slowly
b) quickly

....

An astronaut on a spacecraft suspends a 5.5 kg mass from a newton balance. The reading is 8.8 N.

Q6: The spacecraft might be stationary on the surface of the Earth.

a) True
b) False

....

Q7: The spacecraft might be stationary on the surface of the Moon.

a) True
b) False

....

Q8: The spacecraft might be accelerating away from the surface of the Earth at 1.6 m s^{-2}.

a) True
b) False

....

Q9: The spacecraft might be accelerating away from the surface of the Moon at 1.6 m s^{-2}.

a) True
b) False

....

Q10: The spacecraft might be accelerating in deep space at 1.6 m s^{-2}.

a) True
b) False

....

Q11: Two astronauts awake from a deep sleep onboard a space capsule. They don't know whether they have landed on the surface of a planet or whether they are accelerating in deep space.
Can they perform an experiment inside the capsule to decide which situation they are in?

a) Yes
b) No

....

Q12: A person in freefall is equivalent to a body at rest in deep space since their downwards acceleration exactly cancels out the effect of being in a gravitational field.

a) True
b) False

....

Complete the following statements to form an explanation for gravitational lensing.

Q13: An astronomical body of large mass warps _____ so that it is curved.

....

Q14: The shortest path for _____ from a distant object is now curved and not straight.

....

Q15: The _____ bends round the large mass.

....

Q16: A line on a spacetime diagram which maps a particle's spatial location at every instant in time is called a _____ .

....

Q17: Curved lines on spacetime diagrams correspond to non-inertial frames of reference. This means objects which are _____ .

....

Q18: Which of the following correctly describes the objects' motions?

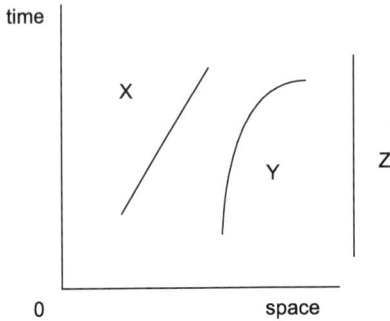

a) X: constant speed, Y: accelerating, Z: stationary
b) X: accelerating, Y: decelerating, Z: constant speed
c) X: constant speed, Y: decelerating, Z: accelerating
d) X: constant speed, Y: decelerating, Z: stationary

....

Q19: Accelerating objects are shown on spacetime diagrams as having world lines of changing _____ .

....

Q20: Which two labels represent simultaneous events?

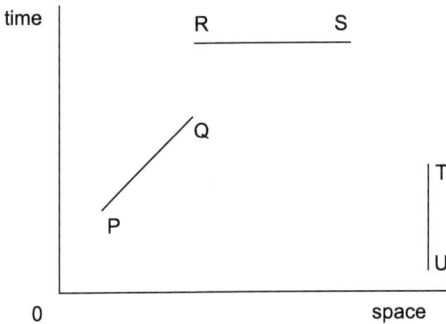

a) T and U
b) R and S
c) P and Q
d) Q and R

....

Q21: Complete the spacetime diagram using the following labels:

The present   v > c   **The future**   v = c   **The past**

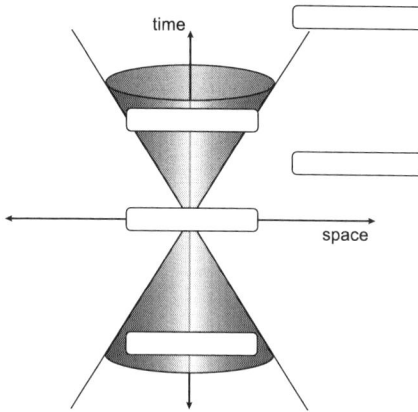

Q22: Which of the following statements about black holes are true?

I   The Schwarzschild radius is the distance from the centre of a black hole to the event horizon.

II   The escape velocity for black holes is greater than the speed of light.

III   A black hole results from the extreme curvature of space due to its compact mass.

a) I only
b) III only
c) I and III only
d) II and III only
e) I, II and III

Q23: Calculate the Schwarzschild radius for a black hole of mass 5.97×10^{31} kg.

Give your answer to *at least* 2 significant figures.

$r_{Schwarzschild} = $ _____ m

Q24: The event horizon is 4.31×10^{0} m from a black hole's singularity. Determine the mass of the black hole.

Give your answer to *at least* 2 significant figures.

$M = $ _____ kg

Topic 7

Stellar physics

Contents

Prerequisite knowledge

- *Irradiance and inverse square law (Traditional or CfE Higher)*

- *Nuclear fusion (Traditional or CfE Higher)*

- *The relationship between the peak wavelength and the temperature of an object (CfE Higher)*

Learning objectives

By the end of this topic you should be able to:

- *state that all stellar objects give out a wide range of wavelengths of electromagnetic radiation but that each object gives out more energy at one particular wavelength;*

- state that the wavelength of this peak wavelength is related to the temperature of the object, with hotter objects having a shorter peak wavelength than cooler objects;

- state that peak wavelengths allow the temperature of stellar objects to be calculated;

- state that hotter objects also emit more radiation per unit surface area at all wavelengths than cooler objects;

- carry out calculations using the fact a star's power per unit area $= \sigma T^4$;

- state that the luminosity of a star, in watts, is a measure of the total power the star emits i.e. the total energy emitted per second;

- state that the luminosity of a star depends on its radius and surface temperature;

- carry out calculations using the equation for luminosity $L = 4\pi r^2 \sigma T^4$;

- state that the apparent brightness of a star is the amount of energy per second reaching a detector per unit area;

- state that the apparent brightness of a star depends on its luminosity and its distance from the observer;

- carry out calculations using the equation for apparent brightness $b = \frac{L}{4\pi r^2}$;

- state that the mass of a new star determines its luminosity and surface temperature;

- state that a Hertzsprung-Russell (H-R) diagram is a plot of luminosity versus the surface temperature of stars and that the surface temperature scale is in descending order;

- identify the long diagonal band on a H-R diagram as the main sequence and state that the Sun is on the main sequence;

- state that the main sequence stars are in their long lived stable phase where they are fusing hydrogen into helium in their cores;

- identify the higher luminosity and lower temperature stars lying to the right of the main sequence as the red giants and red supergiants;

- identify the lower luminosity and higher temperature stars lying to the left of the main sequence as the white dwarfs;

- predict the colour of a star based upon its position on the Hertzsprung-Russell diagram;

- describe how stars produce heat energy using the proton-proton chain reaction;

- describe how stars are formed in terms of the gravitational effects on cold dense interstellar clouds;

- state that the energy released from nuclear fusion in a star results in an outwards thermal pressure;

- *explain how stars on the main sequence are in gravitational equilibrium;*

- *explain why a star's life cycle is determined by its mass;*

- *relate the process of stellar evolution for a star to its path on a H-R diagram.*

7.1   Introduction

In this topic we are going to examine different types of star, such as **red giants** and **white dwarfs**. We will explore what happens to a star when it gets old and we will find out how astronomers use diagrams to keep track of their progression along the life cycle. We will even find out how it can be argued that we are all made of star material.

7.2   Properties of stars

7.2.1   Temperature and colour

A star is a massive body of gas which emits light. The colour of a star is a good guide as to its approximate surface temperature. For example dull red stars are cool and bluish-white stars are very hot.

You will see in Unit 2 that an ideal black body is an object that is both a perfect emitter and absorber of electromagnetic radiation. That is, a black body emits and absorbs all frequencies equally and perfectly. Stars are not ideal black bodies, but treating them as such is a good approximation.

You may remember from Higher that the energy which a star emits is spread over a wide range of wavelengths. Astronomers can analyse a star's intensity-wavelength graph and use the peak wavelength to determine the surface temperature with the equation $\lambda_{max}T = $ constant.

The shorter the wavelength of the peak wavelength, the higher the surface temperature of the star.

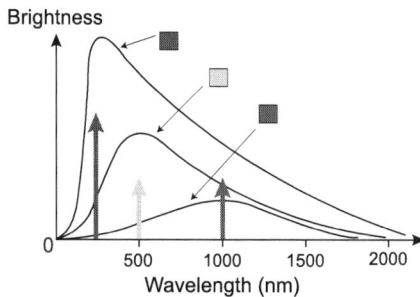

Brightness / Wavelength (nm) graph

Algol (a strong UV source)	
T = 12000 K	λ_m = 250 nm
Sun	
T = 6000 K	λ_m = 500 nm
Proxima Centauri (a strong infrared source)	
T = 3000 K	λ_m = 1000 nm

7.2.2 Spectral class

The relative strengths of particular absorption lines gives the **spectral class** of a star. O is the hottest and M is the coolest type. So the Sun is a fairly middle of the road G star, but the relationship between surface temperature and spectral class is neither linear nor logarithmic.

Class	Temperature (K)	Colour	Notes	Examples
O	≥ 33,000	blue	Most of the electromagnetic radiation given out is ultraviolet	Several stars in the Orion constellation
B	10,000 - 33,000	blue to blue white	These stars are short lived	Rigel
A	7,500 - 10,000	white	This type of star is fairly common in our part of the galaxy	Sirius (one of the brightest stars in the sky) Deneb
F	6,000 - 7,500	yellowish white	Again these star types are common in our part of the galaxy	Capella
G	5,200 - 6,000	yellow	Our own star, the sun is a class G star	The Sun, Polaris (northern pole star)
K	3,700 - 5,200	orange	May be suitable for sustaining life on solar system in their orbit	Arcturus, Aldebaran
M	≤ 3,700	red	The most common of all types of star	Barnard's star

You can use a mnemonic to remember the order. The standard one is "Oh Be a Fine Girl/Guy, Kiss Me", but some may prefer "Oh Boy, An F Grade Kills Me".

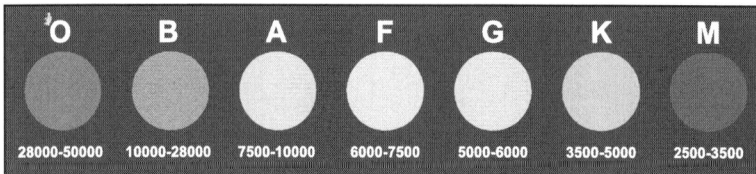

O	B	A	F	G	K	M
28000-50000	10000-28000	7500-10000	6000-7500	5000-6000	3500-5000	2500-3500

The above diagram is clearly not to scale, since the hottest stars are also generally the biggest and brightest. These stars use up their supply of hydrogen very quickly and therefore have shorter lifetimes. In a sense they are a bit like certain movie stars: the brightest largest ones tend to live fast and die young. In fact, all of the stellar properties we shall meet (spectral type, **luminosity**, radius and temperature) depend on a star's mass. In section 7.6 we will see that once a star evolves beyond its normal hydrogen fusing phase, its size and temperature will change, but the life cycle it will follow is also determined by its mass. So for a star, its mass is really important.

O B A F G K M

Decreasing mass →

Decreasing luminosity →

Decreasing temperature →

Decreasing radius →

Increasing lifetime →

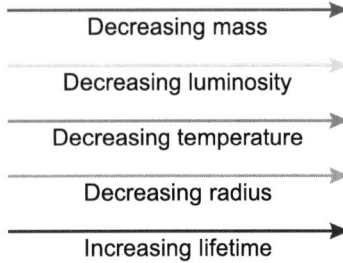

7.2.3 The power of a star

The power output P of a radiating object depends on its temperature and surface area. This relationship is known as the Stefan-Boltzmann Law and is written as follows:

$$P = \sigma A T^4$$

(7.1)

. . . .

where
P is the power output in watts
T is the surface temperature in kelvin
A is the surface area in square metres
σ is the Stefan-Boltzmann constant, which has a value $5.67 \times 10^{-8} W m^{-2} K^{-4}$.

This agrees with everyday observation, since a hot cup of tea cools faster than a lukewarm one. Furthermore, large objects cool faster, since their bigger surface area means they can emit more energy per second than smaller objects.

Using Equation 7.1, a star's power per unit area can be therefore be found from:

$$\text{Power per unit area} = \sigma T^4$$

(7.2)

. . . .

This relationship means that a small increase in a star's surface temperature results in a very large increase in its power output per unit area.

Examples

1.

Vega has a surface temperature of 9602 K. Calculate the power per unit area for Vega.

power per unit area $= \sigma T^4$

power per unit area $= 5.67 \times 10^{-8} \times 9602^4$

power per unit area $= 4.82 \times 10^8 W m^{-2}$

. . . .

2.

The power per unit area of Rigel is $8.30 \times 10^8 W m^{-2}$. Find the surface temperature of Rigel.

power per unit area $= \sigma T^4$

$8.30 \times 10^8 = 5.67 \times 10^{-8} \times T^4$

$T = 11000 K$

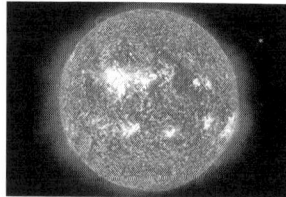

. . . .

7.2.4 Stellar luminosity

The luminosity L of a star is a measure of the total power the star emits from all wavelengths i.e. the total energy emitted per second. We can therefore see from the Stefan-Boltzmann law (Equation 7.1) that the luminosity can be expressed as:

$$L = \sigma A T^4$$

(7.3)

. . . .

Luminosity is measured in watts. For a spherical star of radius r, the surface area is $4\pi r^2$. So stellar luminosity can be calculated using the equation:

$$L = 4\pi r^2 \sigma T^4$$

(7.4)

. . . .

Therefore, the luminosity of a star depends on both its radius and surface temperature. A star can be luminous because it is hot or it is large or both. You may find it helpful to think of the luminosity as being a measure of the true brightness of a star. However, it should be appreciated that a star's total luminosity takes account of all the wavelengths of radiation, such as infrared and microwaves, not just the visible range.

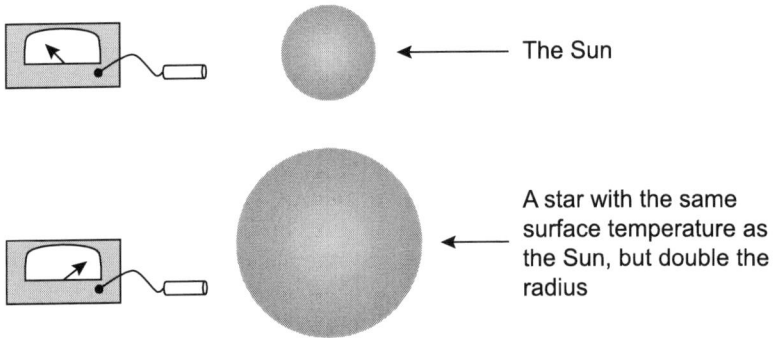

 The Sun

 A star with the same surface temperature as the Sun, but double the radius

The luminosity is proportional to the surface area of the star.

Consider a star which has a radius double that of the Sun. Its surface area would be four times that of the Sun. If this star and the Sun have the same surface temperature, then this star would have a luminosity four times that of the Sun. Considering Equation 7.3 and Equation 7.4, their luminosities can be compared as follows.

$$\frac{L_{star}}{L_{sun}} = \frac{\text{surface area}_{star}}{\text{surface area}_{sun}} = \frac{2^2}{1^2} = 4$$

Star A   Star B   Star C

      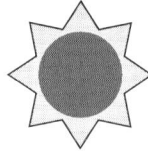

6000 K   12000 K   2000 K

$$L_{Star\ A}$$

$$L_{Star\ B} = \left(\frac{12000}{6000}\right)^4 L_{Star\ A}$$

$$L_{Star\ B} = 2^4\ L_{Star\ A}$$

$$L_{Star\ B} = 16\ L_{Star\ A}$$

$$L_{Star\ C} = \left(\frac{2000}{6000}\right)^4 L_{Star\ A}$$

$$L_{Star\ C} = \left(\frac{1}{3}\right)^4 L_{Star\ A}$$

$$L_{Star\ C} = \frac{1}{81}\ L_{Star\ A}$$

The luminosity is proportional to the fourth power of the surface temperature of the star.

It is worth noting that Equation 7.1 and Equation 7.3 are not provided on the Relationships sheet in the exam, but Equation 7.4 and Equation 7.2 will be included.

Example

The Sun has a surface temperature of 5780 K. The radius of the Sun is 6.955× 10⁸ m. Calculate the luminosity of the Sun.

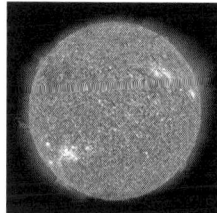

$$L = 4\pi r^2 \sigma T^4$$

$$L = 4\pi \left(6.955 \times 10^8\right)^2 \times 5.67 \times 10^{-8} \times 5780^4$$

$$L = 3.85 \times 10^{26}\,W$$

. . . .

Examples

1.

Betelgeuse has a surface temperature of 3500 K. The luminosity of Betelgeuse is 7.2 × 10^{31} W. Calculate the radius of Betelgeuse.

$$L = 4\pi r^2 \sigma T^4$$
$$7.2 \times 10^{31} = 4\pi r^2 \times 5.67 \times 10^{-8} \times 3500^4$$
$$r = 8.2 \times 10^{11} m$$

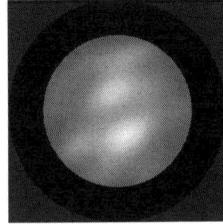

. . . .

2.

Sirius is the brightest star in the night sky. Sirius has a luminosity of 9.9 × 10^{27} W. The radius of Sirius is 1.2 × 10^6 km. Calculate the surface temperature of Sirius.

$$L = 4\pi r^2 \sigma T^4$$
$$9.9 \times 10^{27} = 4\pi (1.2 \times 10^9)^2 \times 5.67 \times 10^{-8} \times T^4$$
$$T = 9900 \ K$$

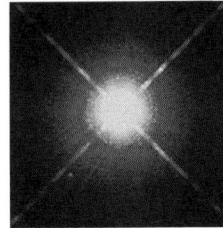

. . . .

3.

Arcturus has a surface temperature of 4300 K and a radius of 1.8 × 10^{10}m. Another star has the same luminosity but a surface temperature of only 3600K. Determine its radius.

$$r^2 T^4 = r^2 T^4$$
$$\left(1.8 \times 10^{10}\right)^2 \times (4300)^4 = r^2 \times (3600)^4$$
$$r^2 = 6.59 \times 10^{20}$$
$$r = 2.6 \times 10^{10} m$$

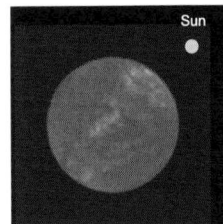

. . . .

7.2.5 Apparent brightness

If you look up at the night sky on a clear night, you will notice that the stars appear to have different levels of brightness. Some appear dimmer than others. Now this may be because they genuinely emit less visible light, but it may also be because they are further away.

For example, consider two stars A and B. Star B emits more visible radiation per second than A. However, the two stars may appear equally bright to you an observer on Earth if star B is more distant than star A. What you are judging with your eye is the **apparent brightness** of the stars.

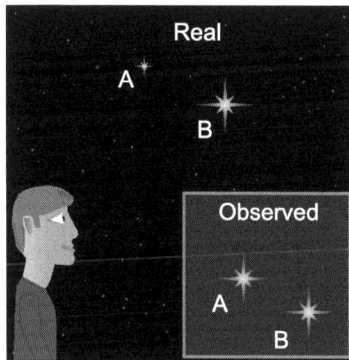

The apparent brightness of a star is a measure of how much electromagnetic radiation actually reaches a detector. Strictly speaking, it takes account of all the wavelengths of the electromagnetic spectrum, not just visible light. The proper astronomical definition for apparent brightness is the amount of energy per second reaching a detector per unit area. It depends upon the star's luminosity and its distance from us. The exact relationship between apparent brightness and distance is called the inverse square law. You may remember it from the Higher course.

Consider the radiation leaving a star and travelling through space. The apparent brightness of the star is the amount of energy per second reaching us per unit area. As you move away from a star, the radiation spreads out to cover the surface of a progressively larger sphere. The area of a sphere is equal to $4\pi r^2$, where r is its radius.

Thus, the amount of energy per unit area received by a detector (the star's apparent brightness) will vary inversely as the square of the star's distance from us. Doubling the distance from a star will quarter the apparent brightness. Tripling the distance will make the apparent brightness become one ninth of the previous value etc.

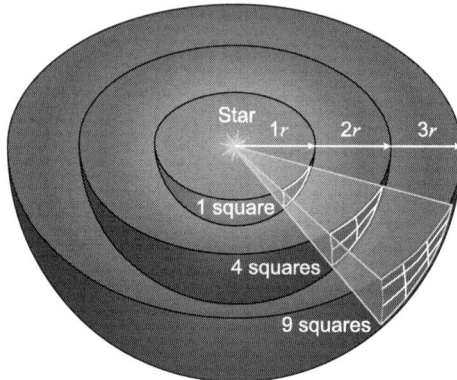

So the apparent brightness b of a star can be found using the equation

$$b = \frac{L}{4\pi r^2}$$

(7.5)

....

where L is the luminosity and r is the distance between the observer and the star.

7.2.6 Knowing your symbols

The next diagram summarises the difference between luminosity and apparent brightness. Make sure you do not confuse the symbol r in the equation for luminosity ($L = 4\pi r^2 \sigma T^4$) with the symbol r in the equation for apparent brightness ($b = \frac{L}{4\pi r^2}$). In the luminosity equation, r stands for the radius of the star. In the apparent brightness equation, r stands for the distance from the star.

Luminosity is the total power a star emits into space
i.e. the total energy emitted per second

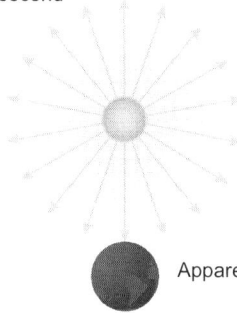

Apparent brightness

This drawing is not to scale

7.2.7 Summary of properties

Astronomers use a star's apparent brightness and distance to find its luminosity. They can then use this and its temperature to determine its size. Indeed all of the properties we have met are interlinked as shown:

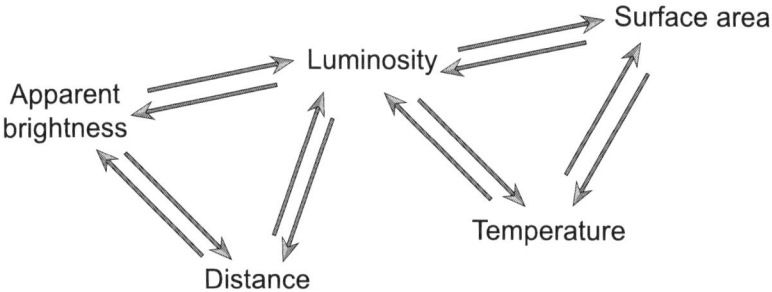

Example

Pollux is a star with a radius of 5.6 × 10^8 m. It is 3.2 × 10^{17} km from Earth and has a surface temperature of 4900 K. Find the apparent brightness of Pollux.

$$L = 4\pi r^2 \sigma T^4$$

$$L = 4\pi \left(5.6 \times 10^8\right)^2 \times 5.67 \times 10^{-8} \times 4900^4$$

$$L = 1.29 \times 10^{26} W$$

$$b = \frac{L}{4\pi r^2}$$

$$b = \frac{1.29 \times 10^{26}}{4\pi \times \left(3.2 \times 10^{20}\right)^2}$$

$$b = 1.0 \times 10^{-16} W m^{-2}$$

. . . .

Useful units in astronomy

You met the definition of a light year in the National 5 course. One light year is the distance that light travels in one year. Therefore, 1 light year can be found from

$$d = vt$$
$$d = 3 \times 10^8 \times 365.25 \times 24 \times 60 \times 60$$
$$d = 9.47 \times 10^{15} m$$

Another useful unit for distance is the Astronomical Unit (AU). The AU is the average distance between the Earth and the Sun, which is about 150 million kilometres. Astronomical units are usually used to measure distances within our solar system and the value is stated on the AH data sheet as

$$1 \text{ AU} = 1 \cdot 5 \times 10^{11} m$$

Similarly, the solar radius ($R_\odot$) is a unit of distance used to express the radius of a star in relation to the current radius of the Sun. The data sheet states that 1 solar radius $= 6 \cdot 955 \times 10^8 m$.

The solar mass ($M_\odot$) is a unit of mass in astronomy that is used to express the mass of a star in relation to that of the Sun. The data sheet states that 1 solar mass $= 2 \cdot 0 \times 10^{30} kg$.

7.3   Star formation

Stars are born in huge clouds of gas and dust called nebulae. These regions are extremely cold at temperatures of about 10 K to 20 K. Higher temperatures would cause the particles to have too much kinetic energy to stay together.

Image Credit: NASA/STScI Digitized Sky Survey/Noel Carboni

Gravitational forces pull the gas and dust into the centre. This accumulation of mass increases the gravitational attraction and results in even more hydrogen gas being pulled in. This leads to an increase in density. Since the gravitational potential energy of the gas is converted to heat, the temperature and pressure at the core also increase. Hydrogen nuclei are positively charged and therefore repel each other, but eventually the temperature of the core will become sufficiently large for the hydrogen nuclei to have enough kinetic energy to be moving fast enough to overcome their electrostatic repulsion and undergo fusion.

Gravitational attraction pulls material inwards,
increasing the temperature and eventually allowing fusion

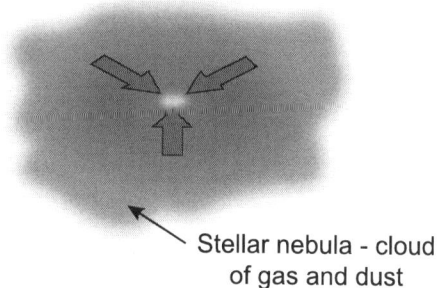

Stellar nebula - cloud
of gas and dust

7.4 Stellar nucleosynthesis and fusion reactions

In this section we are going to look at why stars are so hot and bright- the origin of solar energy. You will remember from the Higher course that nuclear fusion involves nuclei joining together to form a larger nucleus and that this results in a release of energy according to $E = mc^2$. The energy released from nuclear fusion in a star is ultimately released as radiation from its surface.

Stellar nucleosynthesis is the name of the process whereby a star carries out nuclear fusion to produce new elements heavier than hydrogen. A star's mass determines what types of nucleosynthesis occur in its core. This is because heavier stars have a stronger gravitational pull in their cores, leading to higher core temperatures. The smallest stars only convert hydrogen into helium by a process called the proton-proton chain. They can't fuse heavier elements, since they are not sufficiently hot to overcome the repulsive forces between larger nuclei.

The proton-proton chain consists of three stages:

1. Two hydrogen nuclei (^1_1H) undergo fusion to produce a deuterium nucleus (^2_1H), a positron and a neutrino. The positron is annihilated by an electron to produce further energy in the form of gamma ray photons.

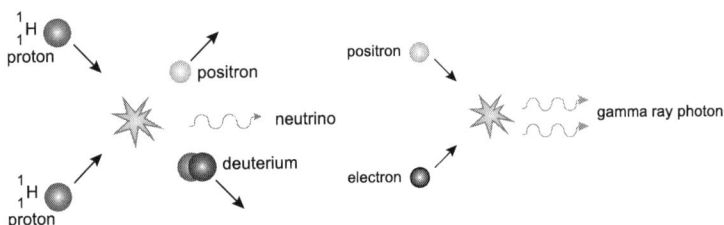

2. The deuterium nucleus (^2_1H) then fuses with a proton to form a helium - 3 nucleus (^3_2He) and a gamma ray photon.

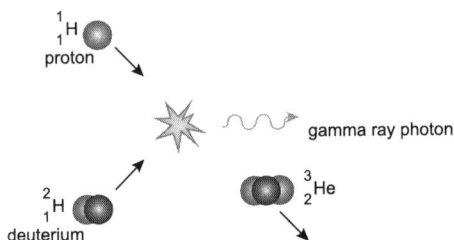

3. Two helium-3 nuclei (^3_2He) fuse to form a helium-4 nucleus (^4_2He) and two free protons, which may cause further proton- proton chains.

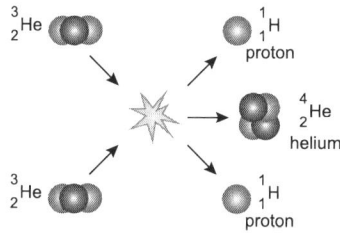

The net effect is to combine 4 protons to form one helium nucleus, with the energy released going into the particles and gamma ray photons produced at each step of the chain.

$$^1_1H + ^1_1H \rightarrow ^2_1H + e^+ + v + \text{energy}$$
$$^1_1H + ^2_1H \rightarrow ^3_2He + \gamma + \text{energy}$$
$$^3_2He + ^3_2He \rightarrow ^4_2He + 2^1_1H + \text{energy}$$

The following is a very brief summary of the proton- proton chain:

1. A proton fuses with another proton to form deuterium.

2. Then deuterium fuses with a proton to form a helium - 3 nucleus.

3. Two helium - 3 nuclei then fuse to form a helium - 4 nucleus.

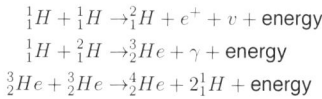

⊙	proton
◉	neutron
∘	neutrino
°	positron
∿∿∿	gamma photon

The proton - proton chain reaction

There is an animation online showing the proton-proton chain reaction, which is the dominant fusion reaction in small stars.

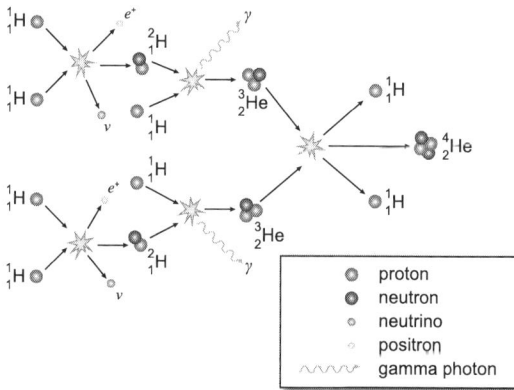

Go online

. . . .

7.5 Hertzsprung-Russell diagrams

Astronomers Hertzsprung and Russell developed the technique of plotting luminosity against surface temperature. Note that both the surface temperature and luminosity scales are logarithmic and that the surface temperature is in descending order. For historical reasons, the lowest surface temperature stars are on the right hand side to match the spectral order OBAFGKM.

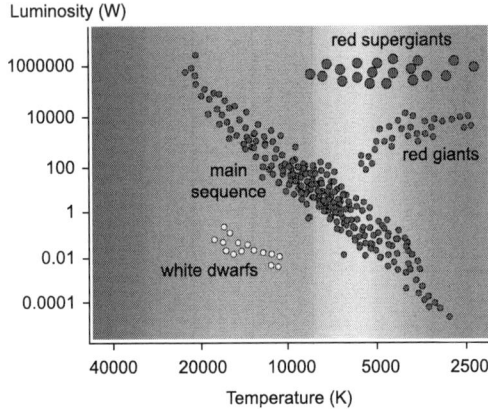

Hot bright stars are shown at the top left of the diagram and cool dim stars are shown at the bottom right.

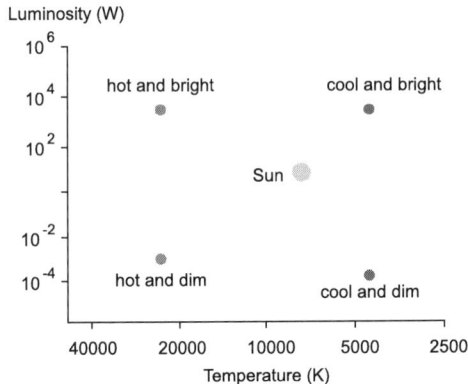

The stars were found to occupy three distinct groups: the **main sequence**, the white dwarfs and the red giants and supergiants.

Luminosity (W)

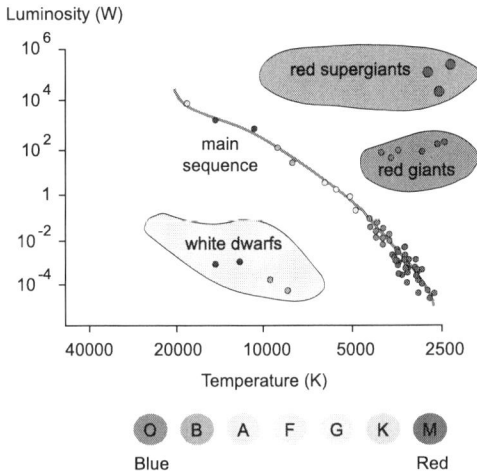

Main sequence

The long diagonal band of stars that runs from the top left to the bottom right is called the main sequence. Most of the stars fall on this curve. Infact about 90% of the stars in our region of the Milky Way belong to the main sequence. Main sequence stars are still in their long-lived stable phase where they are fusing hydrogen into helium in their cores.

The mass of such a new star determines its luminosity and surface temperature. This is because a high mass star has a larger inwards gravitational pull and therefore a higher core temperature. So, the most massive main sequence stars are shown at the upper left. They have a high luminosity and are hot (blue). The least massive ones are shown at the lower right. They are less luminous and are cooler (yellow or red). Since the Sun is neither especially large nor small in mass when compared to the rest of the main sequence, it is neither especially hot, cold, bright or dim. This means that the Sun lies fairly close to the middle of the main sequence.

We will see in section 7.6 that stars begin their life on the main sequence and then evolve to different parts of the H-R diagram. Most of the lifetime of a star is spent on the main sequence.

Red giants and red supergiants

You may recall the luminosity of a star can be found from $L = 4\pi r^2 \sigma T^4$. So stars which have a lower surface temperature and a higher luminosity than the main sequence stars must also have larger radii. These stars are called red giants and **red supergiants**. They are found in the top right corner of the H-R diagram. They are stars which have moved off the main sequence, having started to fuse heavier elements. Red supergiant stars are the brightest of all stars.

White dwarfs

The equation $L = 4\pi r^2 \sigma T^4$ also leads to the conclusion that stars which have a higher surface temperature and a lower luminosity than the main sequence stars must also have smaller radii. These are the white dwarfs. They lie in the bottom left corner of the H-R diagram and are typically about the size of the Earth. White dwarfs are faint hot stars at the end of their lives. All of their fusion reactions have stopped and they are just slowly cooling and dimming to ultimately form a **black dwarf**.

You could argue that the white dwarfs are a bit like old faded Hollywood actors in that they are not as bright or as large as they once were. They have stopped working and are slowly but surely fading away. However, you may contest that the analogy breaks down with regards their hotness!

It is also worth noting that the word white dwarf is a bit of a misnomer, because some are actually blue. They are just named this way since the vast majority are white.

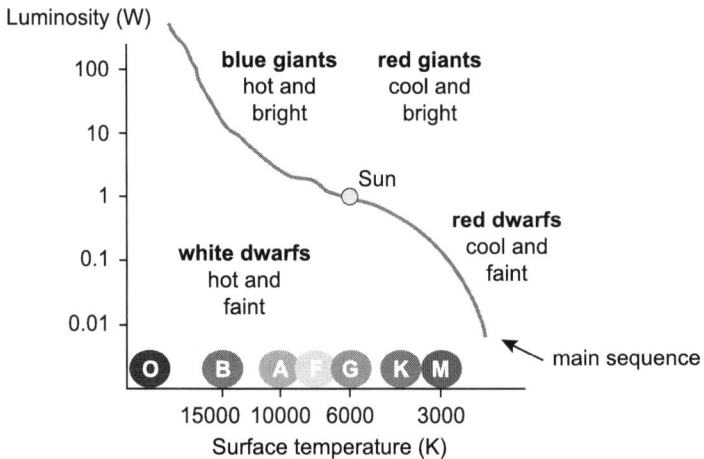

As you can see, the colour of a star can be predicted based upon its position in the Hertzsprung-Russell diagram.

Hertzsprung-Russell diagram matching task

Go online

Q1: Use the labels below to match the correct location on the Hertzsprung- Russell diagram.

Hot bright stars	Cool dim stars	Surface temperature / K	Red giants
Luminosity / W	White dwarfs	Red supergiants	Main sequence

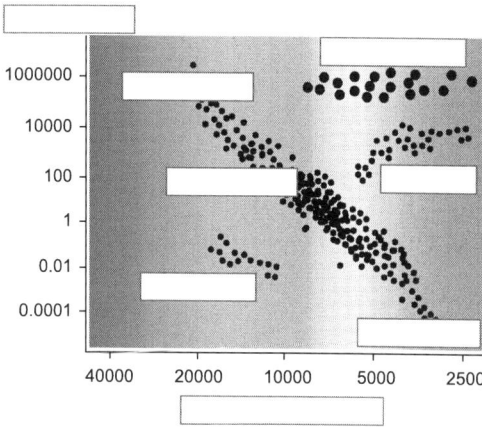

7.6   Stellar evolution and life cycles on H-R diagrams

The mass of a star ranges in size from roughly 0.08 solar masses to 150 solar masses. If an object has a mass less than 0.08 solar masses, then it would be unable to achieve a high enough core temperature to initiate fusion. These "failed stars" are called brown dwarfs. A star cannot have a mass greater than 150 solar masses: the outwards force due to the thermal pressure from fusion would be so large in comparison to the inwards gravitational pull that the extra mass would be driven off into space.

The H-R diagram is a useful tool for tracing the evolution of stars. All main sequence stars will eventually evolve to become either red giants or red supergiants. Which of these they become, the time they spend on the main sequence and their eventual fate also depend on their mass. Even though stars with a large mass have a lot of fuel, they use it up more quickly and therefore don't spend as long on the main sequence. For instance, a star with a mass comparable to that of the Sun will spend about 10 billion years on the main sequence. In comparison, a star that has a mass about 15 times that of the Sun will only belong to the main sequence for approximately 10 million years.

During the stable stage whilst a star is still on the main sequence, the outwards force due to the thermal pressure produced from hydrogen fusion in the core balances the gravitational pull trying to compress it.

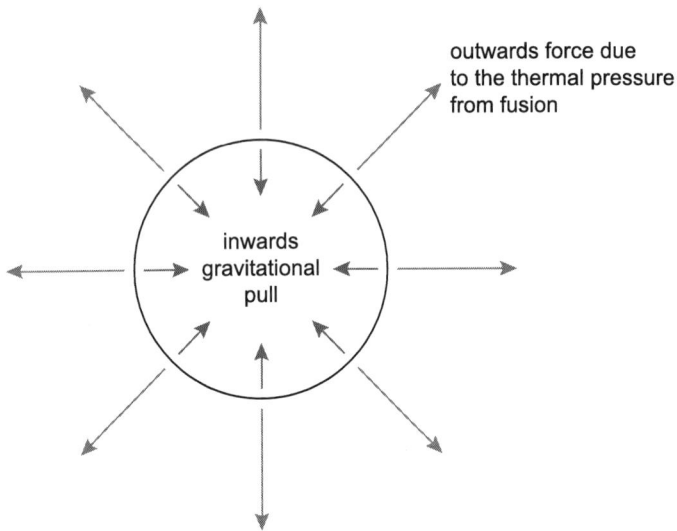

Equilibrium: The outwards force due to thermal pressure from fusion balances the inwards gravitational pull.

Eventually, the hydrogen in the core runs out and nuclear fusion stops. Then, since the outward pressure is removed, the core contracts and heats up. Eventually the temperature is high enough for the core to fuse helium into carbon and oxygen. This releases a huge amount of energy, which pushes the outer layers of the star outwards. As the outer layers expand and cool, the star leaves the main sequence and becomes either a red giant or a red supergiant, depending on its mass.

Death of a low or medium mass star

The core temperature of a low or medium mass star is not sufficiently high to allow the fusion of elements beyond helium. Therefore, once all the helium has been converted into carbon and oxygen, the core once more contracts due to gravitational attraction. The outer layers become more and more unstable as the core contracts. The star pulsates and ejects the outer layers into space as a planetary **nebula**. This leaves behind a dense core-a white dwarf. Remember the white dwarfs lie below the main sequence as they have low luminosities. A white dwarf will then simply cool down and fade away, ultimately forming a black dwarf.

Death of a high mass star

Really massive stars become red supergiants, which can fuse progressively heavier elements all the way up to iron. Once they are depleted, the core collapses in less than a second and a **supernova** explosion occurs. A supernova event can temporarily outshine a whole galaxy. It provides the energy needed to produce all the elements heavier than iron up to uranium. In fact, every human being is made from atoms that were formed in a supernova. So we're all made of star material after all! Shock waves blow off the outer layers, leaving behind either a **neutron star** or a black hole.

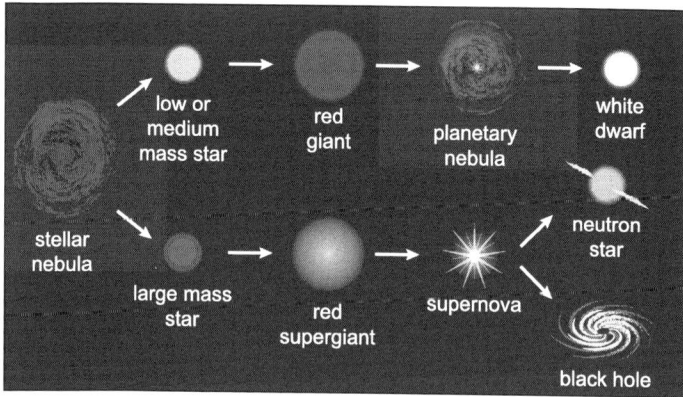

The life cycle of a star

The evolution of a star can be mapped out on a H-R diagram. The one below shows the life cycle of a star of mass equal to that of the Sun. As you can see it starts on the main sequence and then moves diagonally up to the right to become a red giant, before eventually becoming a white dwarf.

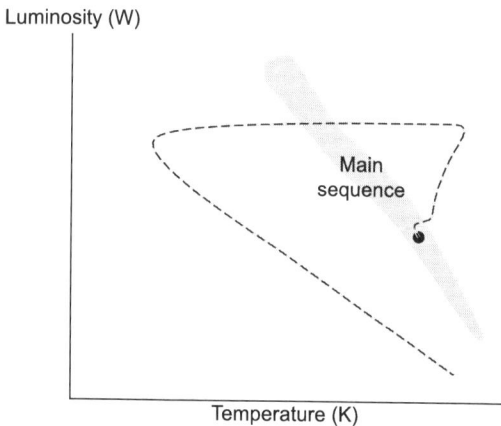

Interactive H-R diagram

There is an interactive version of H-R diagram available online where you will find out more details about Stellar evolution and life cycles.

Go online

....

Stellar matching task

Q2: Match each evolution stage (1 - 6) with their descriptions (A - F).

Go online

1. **Nebula**
2. **Black dwarf**
3. **White dwarf**
4. **Supernova**
5. **Neutron star**
6. **Black hole**

A) The remains of a supernova. It exerts such a strong gravitational pull that light cannot escape.

B) Faint hot stars that lie below and to the left of the main sequence. They are the remnants of a medium or low mass star.

C) A high density small star left over from a supernova.

D) A large cloud of gas and dust from which a star is formed.

E) The remains of a white dwarf after it has cooled.

F) The explosion of a red supergiant.

. . . .

The life cycle of a low or medium mass star

Q3: Place the following statements about the life cycle of a low or medium mass star into the correct sequence 1 - 6.

Go online

- The centre of the nebula's temperature increases.
- Hydrogen runs out in the core and a red giant forms.
- The outer layers of the red giant drift off into space.
- Nuclear fusion starts.
- A white dwarf is formed.
- Nebula contracts due to gravitational attraction.

1. . . .
2. . . .
3. . . .
4. . . .
5. . . .
6. . . .

. . . .

The life cycle of a large mass star

Q4: Place the following statements about the life cycle of a low or medium mass star into the correct sequence 1 - 6.

Go online

- Hydrogen runs out in the core and a red supergiant forms.
- Nuclear fusion starts.
- The core collapses and a supernova explosion occurs.
- A dense neutron star or a black hole is formed.
- Nebula contracts due to gravitational attraction.
- The centre of the nebula's temperature increases.

1. ...
2. ...
3. ...
4. ...
5. ...
6. ...

. . . .

7.7 Summary

Summary

You should now be able to:

- state that all stellar objects give out a wide range of wavelengths of electromagnetic radiation but that each object gives out more energy at one particular wavelength;

- state that the wavelength of this peak wavelength is related to the temperature of the object, with hotter objects having a shorter peak wavelength than cooler objects;

- state that peak wavelengths allow the temperature of stellar objects to be calculated;

- state that hotter objects also emit more radiation per unit surface area at all wavelengths than cooler objects;

- carry out calculations using the fact a star's power per unit area = σT^4;

- state that the luminosity of a star, in watts, is a measure of the total power the star emits i.e. the total energy emitted per second;

Summary continued

- state that the luminosity of a star depends on its radius and surface temperature;

- carry out calculations using the equation for luminosity $L = 4\pi r^2 \sigma T^4$;

- state that the apparent brightness of a star is the amount of energy per second reaching a detector per unit area;

- state that the apparent brightness of a star depends on its luminosity and its distance from the observer;

- carry out calculations using the equation for apparent brightness $b = \frac{L}{4\pi r^2}$;

- state that the mass of a new star determines its luminosity and surface temperature;

- state that a Hertzsprung-Russell (H-R) diagram is a plot of luminosity versus the surface temperature of stars and that the surface temperature scale is in descending order;

- identify the long diagonal band on a H-R diagram as the main sequence and state that the Sun is on the main sequence;

- state that the main sequence stars are in their long lived stable phase where they are fusing hydrogen into helium in their cores;

- identify the higher luminosity and lower temperature stars lying to the right of the main sequence as the red giants and red supergiants;

- identify the lower luminosity and higher temperature stars lying to the left of the main sequence as the white dwarfs;

- predict the colour of a star based upon its position on the Hertzsprung-Russell diagram;

- describe how stars produce heat energy using the proton-proton chain reaction;

- describe how stars are formed in terms of the gravitational effects on cold dense interstellar clouds;

- state that the energy released from nuclear fusion in a star results in an outwards thermal pressure;

- explain how stars on the main sequence are in gravitational equilibrium;

- explain why a star's life cycle is determined by its mass;

- relate the process of stellar evolution for a star to its path on a H-R diagram.

7.8 Extended information

Web links

There are web links available online exploring the subject further.

....

7.9 Assessment

End of topic 7 test

The following test contains questions covering the work from this topic.

Go online

The following data should be used when required:

Gravitational acceleration on Earth g	$9.8\ m\ s^{-2}$

Q5: The luminosity of a star depends upon:

a) its surface temperature, its radius and the distance from the observer.
b) its surface temperature and the distance from the observer.
c) its radius and the distance from us.
d) only its surface temperature.
e) its surface temperature and its radius.

....

Q6: The apparent brightness of a star depends on distance from the star.

a) True
b) False

....

Q7: The apparent brightness of a star depends on surface temperature of the star.

a) True
b) False

....

Q8: Alpha Cassiopeiae has an apparent brightness of 4.4×10^{-9} W m^{-2} and is at a distance of 2.2×10^{18} m from Earth.

Calculate its luminosity.

L = _____ W

....

Q9: The Sun has an apparent brightness of 1370 W m^{-2} and is 1.5×10^{11} m from Earth.

The star Procyon A has an apparent brightness of 1.78×10^{-8} W m^{-2} and is 1.0×10^{17} m from Earth.

How many times bigger is the luminosity of Procyon A relative to that of the Sun?

$L_{Procyon\,A}$ = _____ $\times\ L_{Sun}$

....

Q10: The Sun has an apparent brightness of 1.4×10^3 W m^{-2} and is 1.5×10^{11} m from Earth. The radius of the Sun is 6.96×10^8 m.

Sirius A has an apparent brightness of 1.23×10^{-7} W m^{-2}. The luminosity of Sirius A is 25.4 times that of the Sun.

Determine the distance from Earth to Sirius A.

$r_{Sirius\,A}$ = _____ m

....

Q11: When compared to the other stars, the stars which emit the shortest peak wavelength are:

a) red and cool.
b) red and hot.
c) blue and cool.
d) blue and hot.

....

Q12: The long diagonal band on a Hertzsprung-Russell diagram is made up of stars in their long lived stable phase.

This is called the _____ .

....

Q13: As a star ages it progresses _____ the main sequence.

a) up
b) down

....

Q14: All main sequence stars will eventually become a:

a) red giant / red supergiant.
b) black hole.
c) black dwarf.
d) neutron star.
e) white dwarf.

....

Q15: The Sun is a:

a) white dwarf.
b) neutron star.
c) supernova.
d) red dwarf.
e) main sequence star.

....

Q16: When a star is on the main sequence the outwards force due to the thermal pressure from _____ the inwards gravitational pull.

a) fission is greater than
b) fission balances
c) fusion is greater than
d) fusion balances
e) fusion is less than
f) fission is less than

....

Q17: Which type of star has a high surface temperature but a low luminosity?

....

Q18: White dwarf stars have a higher surface temperature and are more dense than a star on the main sequence.

a) True
b) False

....

Q19: White dwarf stars carry out nuclear fission.

a) True
b) False

....

Q20: White dwarf stars carry out nuclear fusion.

a) True
b) False

. . . .

Q21: As a star moves from the main sequence to become a red giant:

a) its surface temperature decreases and luminosity decreases.
b) its surface temperature decreases and luminosity increases.
c) its surface temperature increases and luminosity decreases.
d) its surface temperature increases and luminosity increases.

. . . .

Q22: Complete the chart using the following words:

Main sequence	**Supernova**	**Black dwarf**
Black hole	**Planetary nebula**	**White dwarf**

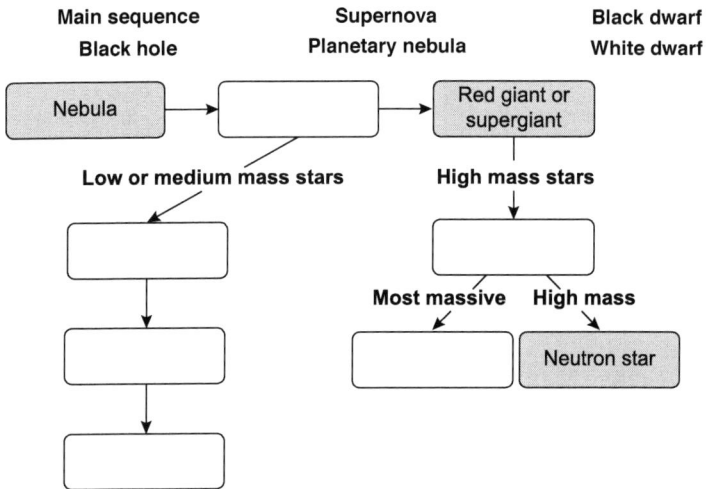

. . . .

Topic 8

End of unit test

End of Unit 1 test

Go online

DATA SHEET

Common Physical Quantities

Quantity	Symbol	Value
Gravitational acceleration on Earth	g	$9.8\ m\ s^{-2}$
Radius of the Earth	r_E	$6.4 \times 10^6\ m$
Mass of the Earth	M_E	$6.0 \times 10^{24}\ kg$
Mass of the Moon	M_M	$7.3 \times 10^{22}\ kg$
Radius of the Moon	r_M	$1.7 \times 10^6\ m$
Mean radius of Moon orbit		$3.84 \times 10^8\ m$
Solar radius	R_S	$6.955 \times 10^8\ m$
Mass of the Sun	M_S	$2.0 \times 10^{30}\ kg$
1 AU		$1.5 \times 10^{11}\ m$
Stefan - Boltzmann constant	σ	$5.67 \times 10^{-8}\ W\ m^{-2}\ K^{-4}$
Universal constant of gravitation	G	$6.67 \times 10^{-11}\ m^3\ kg^{-1}\ s^{-2}$
Planck's constant	h	$6.63 \times 10^{-34}\ J\ s$
Speed of light in a vacuum	c	$3.00 \times 10^8\ m\ s^{-1}$

Q1: A moving particle has a varying force applied to it. Its acceleration in m s^{-2} is given by the expression a = 4.0 t where t is the time in seconds.
Its velocity is 3.0 m s^{-1} at 0 seconds.

Calculate its velocity at 5.0 seconds.

v = _____ m s^{-1}

. . . .

Q2: A mass of 0.420 kg is being rotated by a string in a vertical circle of radius 1.20 m at a constant angular velocity ω rad s^{-1}. The tension in the string when the mass is at the bottom of the circle is 12.0 N.

1. Calculate ω.

   _____ rad s^{-1}

2. Calculate the tension when the mass is at the top of the circle.

   _____ N

3. Calculate the tension when the string is horizontal.

   _____ N

. . . .

Q3: A disc of moment of inertia 40 kg m^2 is made to rotate about an axis through its centre by a torque of T. The disc starts from rest, and after 14 s has kinetic energy 500 J.

1. Calculate the angular velocity after 14 s.

   _ _ _ _ _ _ _ _ _ _ rad s^{-1}

2. Calculate the magnitude of the torque T.

   _ _ _ _ _ _ _ _ _ _ N m

3. The torque is removed when the disc is rotating at 8.0 rad s^{-1}. An opposing torque of 14.5 N m is applied to slow the disc down.
   Calculate the number of revolutions the disc makes after the second torque is applied, before it comes to rest.

   _ _ _ _ _ _ _ _ _ _

. . . .

Q4: A satellite of mass 1060 kg is orbiting the earth at a height of 5.75×10^5 m above the earth's surface.

1. Calculate the gravitational force that the earth exerts on the satellite.

   _ _ _ _ _ _ _ _ _ _ N

2. Calculate the gravitational potential at this height.

   _ _ _ _ _ _ _ _ _ _ J Kg^{-1}

. . . .

Q5: The moon has mass 7.3×10^{22} kg and radius 1.7×10^6 m.

Calculate the escape velocity of the Moon.

_ _ _ _ _ _ _ _ _ _ m s^{-1}

. . . .

Q6: Calculate the Schwarzschild radius for a black hole of mass 2.51×10^{31} kg.

$r_{Schwarzschild}$ = _ _ _ _ _ _ _ _ _ _ m

. . . .

Q7: The surface temperature of the star Wolf 489 is 5030 K. Its luminosity is 3.7×10^{22} W.

What is the radius of Wolf 489?

r = _ _ _ _ _ _ _ _ _ _ m

. . . .

Q8: A star has radius 1.08×10^8 km. It is 1.80×10^{19} m from Earth and its luminosity is 1.27×10^{31} W.

What is the surface temperature of the star?

T = _ _ _ _ _ _ _ _ _ _ K

. . . .

Q9: After hydrogen fusion ends in a star's core, its position on the Hertzsprung-Russell diagram moves towards the:

a) upper right.
b) upper left.
c) lower left.
d) lower right.

. . . .

Q10: Which star is most likely to become a supernova?

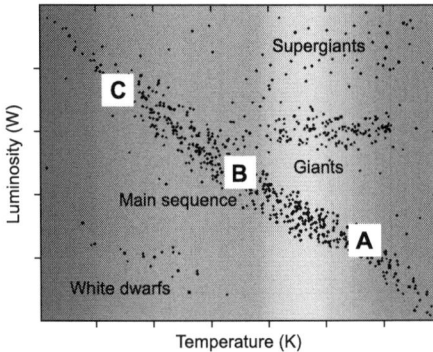

a) A
b) B
c) C

. . . .

Q11: Stars with greater mass have _ _ _ _ _ _ _ _ _ _ life cycles.

a) longer
b) shorter

. . . .

Q12: On a Hertzsprung- Russell diagram, the main sequence stars that lie on the top left of the diagonal line have a:

a) smaller mass and have higher temperatures.
b) larger mass and have higher temperatures.
c) smaller mass and have lower temperatures.
d) larger mass and have lower temperatures.

. . . .

Glossary

Angular acceleration

the rate of change of angular velocity, measured in rad s^{-2}.

Angular displacement

the angle, measured in radians, through which a point or line has been rotated about an axis, in a specified direction.

Angular momentum

the product of the angular velocity of a rotating object and its moment of inertia about the axis of rotation, measured in kg m^2 s^{-1}.

Angular velocity

the rate of change of angular displacement, measured in rad s^{-1}.

Apparent brightness

the amount of energy per second reaching a detector per unit area.

Black body emitter

a perfect absorber and emitter of radiation at all wavelengths of the electromagnetic spectrum.

Black dwarf

the remains of a white dwarf after it has cooled.

Black hole

a region in space from which no matter or radiation can escape, resulting from the extreme curvature of spacetime caused by its compact mass.

Centripetal acceleration

the acceleration of an object moving in a circular path, which is always directed towards the centre of the circle.

Centripetal force

a force acting on an object causing it to move in a circular path.

Conical pendulum

a pendulum consisting of a bob on a string moving in a horizontal circle.

Conservative field

a field in which the work done in moving an object between two points in the field is independent of the path taken.

Equivalence principle

Einstein's principle which states that there is no way of distinguishing between the effects on an observer of a uniform gravitational field and of constant acceleration.

Escape velocity

the minimum speed required for an object to escape from the gravitational field of another object, for example the minimum speed a rocket taking off from Earth would need to escape the Earth's gravitational field.

Event horizon

the boundary of a black hole; no matter or radiation can escape from within the event horizon. Time appears to be frozen at the event horizon of a black hole.

General relativity

the study of non-inertial frames of reference.

Geodesic path

the shortest path between two events in spacetime.

Geostationary satellite

a satellite which orbits the Earth above the equator, with the same periodic time as the Earth's rotation. The satellite remains directly above the same point on the Earth's surface.

Gravitational equilibrium

when the outward force on a star due to thermal pressure from fusion balances the inwards gravitational pull.

Gravitational field

the region of space around an object in which any other object with a mass will have a gravitational force exerted on it by the first object.

Gravitational field strength

the gravitational field strength at a point in a gravitational field is equal to the force acting per unit mass placed at that point in the field.

Gravitational lensing

the bending of light as it passes through curved spacetime.

Gravitational potential

at a particular point in a gravitational field, the gravitational potential is the work done by external forces in bringing a unit mass from infinity to that point.

Gravitational time dilation

the effect whereby time runs more slowly close to an object with a large gravitational field strength.

Gravitational waves

general relativity predicts that the Big Bang created ripples in the curvature of spacetime. Such ripples are also thought to be generated during astronomical events, such as the collision of two black holes.

Hertzsprung-Russell (H-R) diagram

a plot of absolute magnitude or luminosity against spectral class or surface temperature. By convention the surface temperature is in descending order.

Inertial frame of reference

a frame of reference that is stationary or has a constant velocity.

Luminosity

a measure of the total power a star emits i.e. the total energy emitted per second.

Main sequence

stars that are in their long lived stable phase where they are fusing hydrogen into helium in their cores. They form the long diagonal band on a H-R diagram.

Moment of inertia

the moment of inertia of an object about an axis is the sum of mass $\times$ distance from the axis, for all elements of the object.

Nebula

a large cloud of gas and dust from which a star is formed.

Neutron star

a high density small star that is composed almost entirely of neutrons and is leftover from a supernova explosion.

Non-inertial frame of reference

a frame of reference that is accelerating.

Periodic time

the time taken for an object rotating or moving in a circle to complete exactly one revolution.

Planetary nebula

the outer layers of a low or medium mass star that are shed to leave behind a white dwarf.

Precess

the motion of a spinning body in which its axis of rotation changes.

Proton-proton chain

the process whereby a star fuses hydrogen into helium.

Radian

a unit of measurement of angle, where one radian is equivalent to $180/\pi$ degrees. One radian is defined in terms of the angle at the centre of a circle made by the sector of the circle in which the length along the circumference is equal to the radius of the circle

Red giants

high luminosity and low surface temperature stars, lying to the right of the main sequence.

Red supergiant

a large mass star will progress beyond that of a red giant to that of a red supergiant, expanding due to hydrogen depletion. This is the stage before it undergoes a supernova explosion.

Rotational kinetic energy

the kinetic energy of an object due to its rotation about an axis, measured in J. A rotating object still has kinetic energy even if it is not moving from one place to another. The total kinetic energy of a rotating object is the sum of the rotational and translational kinetic energies.

Schwarzschild radius

the distance from the centre of a black hole at which not even light can escape.

Singularity

the point of infinite density at the centre of a black hole to which all mass would collapse.

Spacetime

a representation of four dimensional space, in which the three spatial dimensions and time are blended.

Special relativity

the study of inertial frames of reference.

Spectral class

stars can be put into the classes O,B,A,F,G,K and M, where O is the hottest and M is the coolest.

Stellar nucleosynthesis

process whereby a star carries out nuclear fusion to produce new elements heavier than hydrogen.

Supernova

the explosion of a red supergiant, leaving behind a black hole or a neutron star.

Tangential acceleration

the rate of change of tangential speed, measured in m s^{-2}.

Tangential speed

the speed in m s^{-1} of an object undergoing circular motion, which is always at a tangent to the circle.

Torque

also called the moment or the couple, the torque due to a force is the turning effect of the force, equal to the size of the force times the perpendicular distance from the point where the force is applied to the point of rotation. Torque is measured in units of N m.

Torsion balance

very small forces can be measured using a torsion balance. The forces are applied to the end of a light rod suspended at its centre by a vertical thread. As the forces turn the rod, the restoring torque (turning force) in the thread increases until the turning forces balance.

Universal Law of Gravitation

also known as Newton's law of gravitation, this law states that there is a force of attraction between any two massive objects in the universe. For two point objects with masses m_1 and m_2, placed a distance r apart, the size of the force F is given by the equation:

$$F = \frac{Gm_1m_2}{r^2}$$

Weight

the weight of an object is equal to the force exerted on it by the Earth.

White dwarfs

faint hot stars that are left behind when the outer layers of a red giant drift away. They are the remnants of a medium or low mass star at the end of its life. They lie below and to the left of the main sequence.

Worldline

a line on a spacetime diagram mapping a particle's spatial location at every instant in time.

Hints for activities

Topic 1: Kinematic relationships

Horizontal motion

Hint 1:

List the data you are given in the question

$u = ...\text{m s}^{-1}$
$v = ...\text{m s}^{-1}$
$s = ...\text{m}$
$a = ?$

Once you have the data listed, decide on the appropriate kinematic relationship.

Quiz: Motion in one dimension

Hint 1: First sketch the graph - or see the examples in the section Motion in one dimension.

Hint 2: This is a straight application of the second equation of motion.

Hint 3: This is a straight application of the second equation of motion.

Hint 4: You must choose a positive direction 'up' or 'down' - the initial vertical velocity is up and the final displacement is down - be careful with the sign of the acceleration.

Hint 5: First, work out the displacement of the stone from its starting position. Initial velocity is zero, $a = g$ - use the third equation of motion to find v.

Topic 2: Angular motion

Quiz: Radian measurement

Hint 1: $360° = 2\pi$ radians.

Hint 2: $360° = 2\pi$ radians.

Hint 3: $360° = 2\pi$ radians.

Hint 4: 1 complete rotation = 2π radians

Hint 5: First find out the fraction of the circumference that the object has moved.

Quiz: Angular velocity and angular kinematic relationships

Hint 1: $\omega = 2\pi f$

Hint 2: This is the angular equivalent of a speed, displacement, time calculation.

Hint 3: This is the angular equivalent of a speed, displacement, time calculation.

Hint 4: First work out the change in angular velocity.

Hint 5: The angular displacement after one complete revolution = 2π.

Quiz: Angular velocity and tangential speed

Hint 1: This is a straight application of $v = r\omega$.

Hint 2: This is a straight application of $v = r\omega$.

Hint 3: First work out the circumference of the circle.

Hint 4: First work out the angular deceleration then use $a = r\alpha$.

Hint 5: First work out the angular displacement.

Quiz: Centripetal acceleration

Hint 1: This is a straight application of $a_\perp = r\omega^2$.

Hint 2: See the section titled Centripetal acceleration.

Hint 3: This is a straight application of $a_\perp = \frac{v^2}{r}$

Hint 4: First, work out either ω or $a_\perp = \frac{v^2}{r}$.

Hint 5: Substitute $r = 2r$ in $a_\perp = \frac{v^2}{r}$

Quiz: Horizontal and vertical motion

Hint 1: This is a straight application of $F = \frac{mv^2}{r}$

Hint 2: This is a straight application of $F = \frac{mv^2}{r}$

Hint 3: Consider each statement in turn while referring to $F = \frac{mv^2}{r}$

Hint 4: Calculate the speed for a central force equal to the maximum tension.

Hint 5: At what point is the tension in the string greatest? See the section on vertical motion.

Quiz: Conical pendulum and cornering

Hint 1: See the section titled Conical pendulum.

Hint 2: See the section titled Conical pendulum for the derivation of the relationship

$$\cos\theta = \frac{g}{l\omega^2}$$

. Remember to use SI units.

Hint 3: Start from

$$\cos \theta = \frac{g}{l\omega^2}$$

and remember

$$\omega = 2\pi f$$

Hint 4: See the section titled Cars cornering

Hint 5: See the activity banked corners.

Topic 3: Rotational dynamics

Quiz: Torques

Hint 1: Consider the units on both sides of the relationship $T = Fr$.

Hint 2: This is a straight application of $T = Fr$.

Hint 3: This is a straight application of $T = Fr$.

Hint 4: First, calculate the component of the force perpendicular to the radius.

Hint 5: First calculate the torque and use this to find the component of the force perpendicular to the radius.

Topic 4: Angular momentum

Quiz: Conservation of angular momentum

Hint 1: This is a straight application of $L = I\omega$.

Hint 2: Use the relationship given to find the moment of inertia of the disc. Then apply $L = I\omega$.

Hint 3: Angular momentum is conserved. What effect does adding the clay have on the total moment of inertia?

Quiz: Angular momentum and rotational kinetic energy

Hint 1: Use the relationship given to find the moment of inertia of the sheet. Then apply rotational $E_k = \frac{1}{2}I\omega^2$.

Hint 2: First, work out the value of ω.

Topic 5: Gravitation

Quiz: Gravitational force

Hint 1: This is a straight application of $F = \frac{Gm_1m_2}{r^2}$

Hint 2: This is an example of Newton's Third Law which is sometimes stated as "To every action there is an equal an opposite reaction."

Hint 3: Apply $\text{Weight} = \frac{Gm_1m_2}{r^2}$

Hint 4: What happens to the distance between the object and the centre of the planet?

Hint 5: Consider the weight of a mass of 1 kg on the surface of Venus

Quiz: Gravitational fields

Hint 1: This is a straight application of $g = \frac{GM}{r^2}$

Hint 2: Think of Newton's Second Law!

Hint 3: The gravitational strength is halved - this means the distance from the centre of the planet is increased by a factor $\sqrt{2}$ - see the relationship $g = \frac{GM}{r^2}$

Hint 4: The two gravitational forces must be equal and opposite.

Hint 5: This is a straight application of $g = \frac{GM}{r^2}$

Quiz: Satellite motion

Hint 1: This is a straight application of $v = \sqrt{\frac{GM_E}{r}}$.

Hint 2: Consider the options in terms of the relationships in the section titled Satellite motion.

Hint 3: See the section titled Geostationary satellites.

Hint 4: First, use the relationship $E_p = -\frac{Gm_1m_2}{r}$ to figure out what happens to the potential energy - notice the negative sign!!.
Re kinetic energy, consider the relationship $v = \sqrt{\frac{GM_E}{r}}$; what happens to v when r decreases?

Hint 5: Rearrange the relationship at the end of the section titled Geostationary satellites.

Quiz: Gravitational potential

Hint 1: Consider the relationship

$$V = -\frac{GM}{r}$$

Hint 2: This is a straight application of

$$V = -\frac{GM}{r}$$

Hint 3: This is an application of $V = -\frac{GM}{r}$.
Remember r is the distance from the centre of the Earth.

Hint 4: This is a straight application of $E_P = -\frac{Gm_1m_2}{r}$.

Hint 5: Consider the options in the context of the relationship $E_P = -\frac{Gm_1m_2}{r}$.

Quiz: Escape velocity

Hint 1: See the section titled Escape velocity.

Hint 2: Consider the relationship $v = \sqrt{\frac{2GM_E}{r_E}}$.

Hint 3: This is an application of the relationship $v = \sqrt{\frac{2GM_{\text{planet}}}{r_{\text{planet}}}}$.

Answers to questions and activities

1 Kinematic relationships

Horizontal motion (page 8)

Expected answer

List the data you are given in the question

$u = 12.0$ m s^{-1}

$v = 0$ m s^{-1}

$s = 30.0$ m

$a = ?$

The appropriate kinematic relationship is $v^2 = u^2 + 2as$.

Putting the values into this equation

$$v^2 = u^2 + 2as$$
$$\therefore 0^2 = 12.0^2 + (2 \times a \times 30.0)$$
$$\therefore 0 = 144 + 60a$$
$$\therefore -60a = 144$$
$$\therefore a = -\frac{144}{60}$$
$$\therefore a = -2.40 \text{ m s}^{-2}$$

So to stop the car in exactly 10.0 m, the car must have an acceleration of -2.40 m s^{-2}, equivalent to a deceleration of 2.40 m s^{-2}.

Quiz: Motion in one dimension (page 10)

Q1: a) increases with time.

Q2: d) 45.0 m

Q3: c) 5.05 s.

Q4: d) 13 m s^{-1}

Q5: c) 18.8 m s^{-1}

End of topic 1 test (page 26)

Q6: a = 1.5 m s^{-2}

Q7: Stopping distance = 41 m

Q8: t = 3.3 s

Q9: t = 1.8 s

Q10: The instantaneous acceleration of a body can be found by calculating the **gradient** of the tangent to the velocity-time graph.

Q11:

The displacement can be found from a velocity-time graph by determining the area under the graph. This process is equivalent to **integrating/integration** between limits.

Q12:

$$\frac{dv}{dt} = a$$
$$\int_0^v dv = \int_0^5 4.0t - 1.2 \quad dt$$
$$[v]_0^v = \left[\frac{4.0t^2}{2} - 1.20t\right]_0^5$$
$$v - 0 = \left(\left(\frac{4.0 \times 5^2}{2}\right) - (1.2 \times 5)\right) - 0$$
$$v = 44 \text{ m s}^{-1}$$

Q13:

$$\frac{dv}{dt} = a$$
$$\int_5^v dv = \int_0^3 4.2t - 0.6 \quad dt$$
$$[v]_5^v = \left[\frac{4.2t^2}{2} - 0.6t\right]_0^3$$
$$v - 5 = \left(\frac{4.2 \times (3)^2}{2} - (0.6 \times 3)\right) - 0$$
$$v = 22.1 \text{ m s}^{-1}$$

Q14:

$$\frac{ds}{dt} = v$$
$$\int_{5.0}^s ds = \int_0^{2.0} 9.0t^2 - 0.5t \quad dt$$
$$[s]_{5.0}^s = \left[\frac{9.0t^3}{3} - \frac{0.5}{2}t^2\right]_0^{2.0}$$
$$s - 5.0 = \left(\frac{9.0 \times (2.0)^3}{3} - \frac{0.5 \times (2.0)^2}{2}\right) - 0$$
$$s = 28 \text{ m}$$

2 Angular motion

Quiz: Radian measurement (page 33)

Q1: c) 2.53 rad

Q2: c) 68.75°

Q3: d) $2\pi/_3$ rad

Q4: d) 6π rad

Q5: a) 0.208 rad

Orbits of the planets (page 35)

Expected answer

Planet	Orbit radius (m)	Period (days)	Period (s)	Angular velocity (rad s^{-1})
Mercury	5.79×10^{10}	88.0	7.60×10^6	8.27×10^{-7}
Venus	1.08×10^{11}	225	1.94×10^7	3.24×10^{-7}
Earth	1.49×10^{11}	365	3.15×10^7	1.99×10^{-7}

Quiz: Angular velocity and angular kinematic relationships (page 39)

Q6: e) 50.3 rad s^{-1}

Q7: a) 0.45π rad s^{-1}

Q8: d) 60.0 rad

Q9: b) 0.70 rad s^{-2}

Q10: c) 7.9 s

Quiz: Angular velocity and tangential speed (page 44)

Q11: c) 4.80 m s^{-1}

Q12: c) 1.60 m

Q13: e) 4.58 m s^{-1}

Q14: a) 0.108 m s^{-2}

Q15: b) 3.2 m

Quiz: Centripetal acceleration (page 50)

Q16: c) 8.4 m s^{-2}

Q17: b) The centripetal acceleration is always directed towards the centre of the circle.

Q18: b) 5.6 m s^{-2}

Q19: e) 140 m s^{-2}

Q20: a) The centripetal acceleration halves in value.

Motion in a vertical circle (page 55)

Expected answer

1. The rope is most likely to go slack when the mass is at the top of the circle. Compare Equation 2.12 and Equation 2.13 to understand why.

2. At the top of the circle, Equation 2.12 tells us

$$T_{top} = mr\omega^2 - mg$$

Expressing this in terms of the tangential speed v,

$$T_{top} = \frac{mv^2}{r} - mg$$

When the rope goes slack, the tension in it must have dropped to zero, so

$$0 = \frac{mv^2}{r} - mg$$
$$\therefore \frac{mv^2}{r} = mg$$
$$\therefore v^2 = gr$$
$$\therefore v = \sqrt{gr}$$
$$\therefore v = \sqrt{9.8 \times 0.80}$$
$$\therefore v = 2.8 \text{ m s}^{-1}$$

Quiz: Horizontal and vertical motion (page 56)

Q21: c) 40 N

Q22: c) 6.0 rad s^{-1}

Q23: d) If the radius of the circle is increased, the centripetal force decreases.

Q24: b) 14.4 m s^{-1}

Q25: a) When the mass is at the bottom of the circle.

Quiz: Conical pendulum and cornering (page 64)

Q26: d) $T \times \sin \theta$

Q27: c) $57°$

Q28: e) $2\pi \sqrt{l \cos \theta / g}$

Q29: a) centripetal force is provided by the frictional force.

Q30: d) a component of the normal reaction force contributes to the centripetal force.

End of topic 2 test (page 66)

Q31: 1.15

Q32: Time taken = 1.80 s

Q33: Angular velocity = 3.03 $\times$ 10^{-6} rad s-1

Q34: t = 1.53 s

Q35: Angular deceleration = 4.20 rad s^{-2}

Q36: Speed = 0.691 m s^{-1}

Q37: Average angular acceleration = 0.075 rad s^{-2}

Q38: Centripetal acceleration = 5.31 m s^{-2}

Q39: Radius = 8.6 cm

Q40: Centripetal force = 23.6 N

Q41: Minimum radius = 6.3 m

Q42:

1. Tension = 3.7 N
2. Angular velocity = 3.1 rad s^{-1}

Q43: Maximum speed = 14.0 m s^{-1}

Q44: Maximum speed = 18.5 m s^{-1}

3 Rotational dynamics

Quiz: Torques (page 73)

Q1: e) N m

Q2: d) 49 N m

Q3: a) 3.75 m

Q4: c) 8.46 N m

Q5: d) 111 N

Torque and static equilibrium (page 77)

Q6: $(450 \times 4) = (600 \times$ distance$)$
Distance = 3 m on the right.

Q7: $(450 \times 3) = (600 \times$ distance$)$
Distance = 2.25 m on the right.

Q8: $(450 \times 3) + (480 \times 2) = (600 \times$ distance$)$
Distance = 3.85 m on the right.

Q9: $(450 \times 3) + (480 \times 2) = (360 \times 4) + (600 \times$ distance$)$
Distance = 1.45 m on the right.

Combinations of rotating bodies (page 84)

Q10: 0.16 kg m^2

Q11: 20 cm from the centre

Q12: 0.40 kg m^2

Q13: 0.43 kg m^2

Q14: 28 cm

End of topic 3 test (page 86)

Q15: Moment = 24.9 Nm

Q16: Angular acceleration = 3.8 rad s^{-2}

Q17: I = 0.25 kg m^2

Q18: I = 4.18 kg m^2

Q19: I = 0.12 kg m^2

Q20: The moment of inertia of an object is a measure of its resistance to **angular** acceleration about a given axis.

The moment of inertia of an object about an axis depends on the **mass** of the object, and the distribution of the **mass** about the axis.

Q21: $I = 6.0 \times 10^{-5}$ kg m^2

Q22: m = 15.0 g

Q23: r = 0.951 cm

4 Angular momentum

Quiz: Conservation of angular momentum (page 93)

Q1: d) 7.20 kg m^2 s^{-1}

Q2: b) 0.25 kg m^2 s^{-1}

Q3: a) The turntable would slow down.

Quiz: Angular momentum and rotational kinetic energy (page 96)

Q4: c) 0.015 J

Q5: e) 187 J

End of topic 4 test (page 98)

Q6: Rotational E_K = 0.235 J

Q7: ω = 3.43 rad s^{-1}

Q8: ω = 3.5 rad s^{-1}

Q9:

1. ω = 4.5 rad s^{-1}
2. Total E_K = 76.5 J

5 Gravitation

Quiz: Gravitational force (page 109)

Q1: d) 1.0×10^{-10} N

Q2: a) $F_S = F_E$

Q3: d) 8.4 N

Q4: b) Its mass remains constant but its weight decreases.

Q5: e) 8.87 m s^{-2}

Quiz: Gravitational fields (page 114)

Q6: d) 26.5 N kg^{-1}

Q7: a) m s^{-2}

Q8: a) 7.1×10^5 m

Q9: c) 3.0 m from P

Q10: d) 1.03×10^{26} kg

Quiz: Satellite motion (page 120)

Q11: b) 7700 m s^{-1}

Q12: e) The period of a satellite orbiting the Earth depends on the mass of the satellite.

Q13: c) 3100 m s^{-1}

Q14: d) its potential energy decreases but its kinetic energy increases.

Q15: d) $m_p = \dfrac{4\pi^2 r^3}{GT^2}$

Quiz: Gravitational potential (page 125)

Q16: c) the distance of A from the centre of the Earth.

Q17: c) -2.9×10^6 J kg^{-1}

Q18: c) -6.0×10^7 J kg^{-1}

Q19: d) -4.9×10^{10} J

Q20: a) The satellite has moved closer to the Earth.

Quiz: Escape velocity (page 129)

Q21: e) escape from the Earth's gravitational field.

Q22: b) the mass and radius of the Earth.

Q23: c) 5020 m s^{-1}

End of topic 5 test (page 131)

Q24: F = 2.16 $\times$ 10^{20}

Q25: F = 8.09 $\times$ 10^{-10}

Q26: Weight on Neptune = 74.2 N

Q27: g = 8.9 m s^{-2}

Q28: 3.8 N

Q29: 2.75 N kg^{-1}

Q30: 0.378 N kg^{-1}

Q31: 7.29 $\times$ 10^{-3} N kg^{-1}

Q32: V = -3.2 $\times$ 10^{7} J kg -1

Q33: E$_p$ = -7.2 $\times$ 10^{13}

Q34: V = -2.6 $\times$ 10^{7}

Q35: E$_p$ = -3.2 $\times$ 10^{11}

Q36: 2.6 $\times$ 10^{5} m

Q37: 1.87 $\times$ 10^{4} m s^{-1}

Q38: 3.27 $\times$ 10^{30} kg

6 General relativity and spacetime

End of topic 6 test (page 157)

Q1: b) non-inertial

Q2: c) accelerating

Q3: equivalence principle

Q4: a) slowly

Q5: a) slowly

Q6: b) False

Q7: a) True

Q8: b) False

Q9: b) False

Q10: a) True

Q11: b) No

Q12: a) True

Q13: spacetime

Q14: light

Q15: light

Q16: worldline

Q17: accelerating

Q18: a) X: constant speed, Y: accelerating, Z: stationary

Q19: gradient

Q20: b) R and S

Q21:

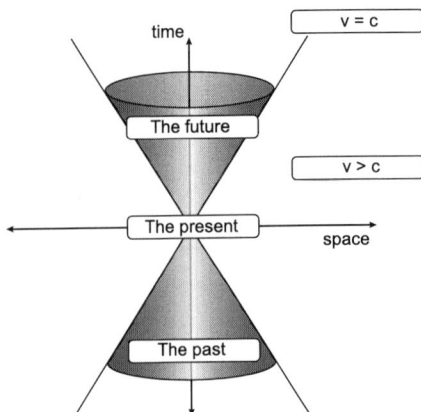

Q22: e) I, II and III

Q23:

$$r_{Schwarzschild} = \frac{2GM}{c^2}$$

$$r_{Schwarzschild} = \frac{2 \times 6.67 \times 10^{-11} \times 5.97 \times 10^{31}}{(3 \times 10^8)^2}$$

$$r_{Schwarzschild} = 8.85 \times 10^4 \text{ m}$$

Q24:

$$r_{Schwarzschild} = \frac{2GM}{c^2}$$

$$4.31 \times 10^8 = \frac{2 \times 6.67 \times 10^{-11} \times M}{(3 \times 10^8)^2}$$

$$M = 2.91 \times 10^{35} \text{ kg}$$

7 Stellar physics

Hertzsprung-Russell diagram matching task (page 182)

Q1:

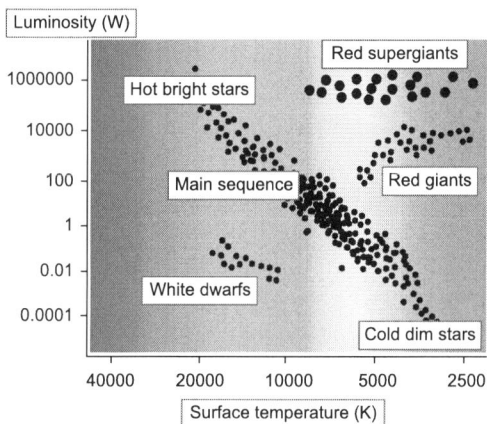

Stellar matching task (page 186)

Q2: 1D; 2E; 3B; 4F; 5C; 6A

The life cycle of a low or medium mass star (page 186)

Q3:

1. Nebula contracts due to gravitational attraction.
2. The centre of the nebula's temperature increases.
3. Nuclear fusion starts.
4. Hydrogen runs out in the core and a red giant forms.
5. The outer layers of the red giant drift off into space.
6. A white dwarf is formed.

The life cycle of a large mass star (page 187)

Q4:

1. Nebula contracts due to gravitational attraction.
2. The centre of the nebula's temperature increases.
3. Nuclear fusion starts.

4. Hydrogen runs out in the core and a red supergiant forms.

5. The core collapses and a supernova explosion occurs.

6. A dense neutron star or a black hole is formed.

End of topic 7 test (page 189)

Q5: e) its surface temperature and its radius.

Q6: a) True

Q7: a) True

Q8: $L = 2.7 \times 10^{29}$

Q9: $L_{Procyon\,A} = 5.8 \times L_{Sun}$

Q10: $r_{Sirius\,A} = 8.1 \times 10^{16}$ m

Q11: d) blue and hot.

Q12: This is called the **main sequence**.

Q13: a) up

Q14: a) red giant / red supergiant.

Q15: e) main sequence star.

Q16: d) fusion balances

Q17: White dwarf

Q18: a) True

Q19: b) False

Q20: b) False

Q21: b) its surface temperature decreases and luminosity increases.

Q22:

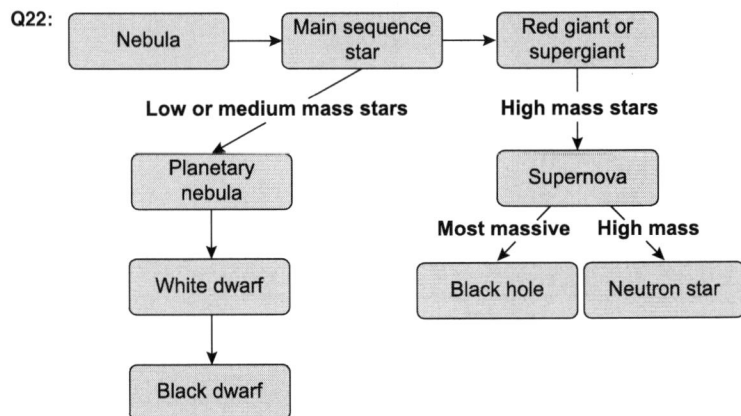

8 End of unit test

End of Unit 1 test (page 194)

Q1: $v = 53$ m s^{-1}

Q2:

1. 3.96 rad s^{-1}
2. 3.79 N
3. 7.90 N

Q3:

1. 5.00 rad s^{-1}
2. 14 N m
3. 14

Q4:

1. 8.7×10^3 N
2. -5.7×10^7 J Kg^{-1}

Q5: 2.4×10^3 m s^{-1}

Q6: $r_{Schwarzschild} = 3.72 \times 10^4$ m

Q7: $r = 9.0 \times 10^6$ m

Q8: $T = 6250$ K

Q9: a) upper right.

Q10: c) C

Q11: b) shorter

Q12: b) larger mass and have higher temperatures.